T0212864

Lecture Notes in Management and Industrial Engineering

Series editor

Adolfo López-Paredes, Valladolid, Spain

This bookseries provides a means for the dissemination of current theoretical and applied research in the areas of Industrial Engineering & Engineering Management. The latest methodological and computational advances that both researchers and practitioners can widely apply to solve new and classical problems in industries and organizations constitute a growing source of publications written for and by our readership.

The aim of this bookseries is to facilitate the dissemination of current research in the following topics:

- Strategy and Enterpreneurship
- Operations Research, Modelling and Simulation
- Logistics, Production and Information Systems
- Quality Management
- Product Management
- Sustainability and Ecoefficiency
- Industrial Marketing and Consumer Behavior
- Knowledge and Project Management
- Risk Management
- Service Systems
- Healthcare Management
- Human Factors and Ergonomics
- Emergencies and Disaster Management
- Education

More information about this series at http://www.springer.com/series/11786

João M. Fernandes · Ricardo J. Machado

Requirements in Engineering Projects

João M. Fernandes
Departamento de Informática
Universidade do Minho
Braga
Portugal

Ricardo J. Machado
Departamento de Sistemas de Informação
Universidade do Minho
Guimarães
Portugal

The Authors retain the sole right to use 16 illustrations commissioned specially for this Work, to which the artist retains the copyright. These illustrations are 1.1, 1.2, 2.1, 2.2, 3.1, 3.2, 4.1, 4.2, 5.1, 5.2, 6.1, 6.2, 7.1, 7.2, 8.1, and 8.2.
Figure 4.3 appears courtesy of Springer, with kind permission.
Wiley is thanked for granting permission to use material from Naveda and Seidman. IEEE Computer Society real-world software engineering problems: A self-study guide for today's software professional. Copyright 2006, by IEEE Computer Society. Published by John Wiley & Sons, Inc., Hoboken, New Jersey, for the exercises 2.4, 3.1, 3.2, 3.3, 3.4, 4.4, 4.6, 5.1, 5.2, 5.3, 6.1, 6.2, 7.1, and 7.2 that appear in this book.

ISSN 2198-0772   ISSN 2198-0780 (electronic)
Lecture Notes in Management and Industrial Engineering
ISBN 978-3-319-36818-4   ISBN 978-3-319-18597-2 (eBook)
DOI 10.1007/978-3-319-18597-2

Springer Cham Heidelberg New York Dordrecht London

Printed on acid-free paper

Springer International Publishing AG Switzerland is part of Springer Science+Business Media (www.springer.com)

*I dedicate this book to my children,
Constança and Gonçalo, and my beloved
wife Raquel.*

João M. Fernandes

*To my dear wife Sofia and my children Inês
and Tiago for their limitless patience,
understanding and support. To my
parents for the education they gave
me as a foundation to my life.*

Ricardo J. Machado

Foreword

Software artefacts belong to the most complicated things ever constructed by humankind. Their complexity has been growing ever since the first computers emerged in the 1940s. Think of the software of the Large Hadron Collider (LHC) and the Boeing 787, but also SAP's enterprise software and the Windows operating system. These artefacts are much more complicated than bridges and skyscrapers. In fact, software forms an integral part of the most complex artefacts built by humans. Software systems may comprise hundreds of millions of program statements, written by thousands of different programmers, spanning several decades. Their complexity surpasses the comprehensive abilities of any single, individual human being. Accordingly, we have become totally dependent on complex software artefacts. Communication, production, distribution, healthcare, transportation, education, entertainment, government, and trade all increasingly rely on "Big Software". Unfortunately, we only recognise our dependency on software when it fails. Malfunctioning information systems of the Dutch police force and the Dutch tax authority, outages of electronic payment and banking systems, increasing downtime of high-tech systems, unusable phones after updates, failing railway systems, and tunnel closures due to software errors illustrate the importance of good software.

The book "Requirements in Engineering Projects" by João Fernandes and Ricardo Machado addresses the engineering challenges mentioned above. They recognise that engineering is a discipline that is intimately related to the concept of project. The authors follow a multidisciplinary approach. Indeed engineering is not about a new notation or lifecycle model. The challenge is to define the interface of a system with its environment in the best way possible. Both natural language and graphical models are considered to describe requirements. Topics related to requirements engineering, in particular elicitation, negotiation, prioritisation, and documentation, are considered in depth. Classical software engineering books only cover requirements superficially and other books on requirements tend to be too abstract. This book provides an excellent mix between practical guidance, focus, and completeness. As a result it is directly useable by practitioners.

After an introduction to software engineering, the book "dives" into the world of requirements. Next to defining the basic concepts of requirements and the principles of requirements engineering, the authors focus on requirements elicitation, negotiation, and prioritisation. Before providing an overview of modelling approaches, recommendations for writing requirements in English are given. The authors did an excellent job in making the book engaging and appealing for a larger audience. The book uses quotes and very nice illustrations. These serve as a catalyst for developing new views on engineering, and make the book a joy to read.

Today, many people talk about "Big Data". However, we should not forget about "Big Software" and "Big Systems". Fortunately, this book provides a concrete approach to requirements in engineering projects. It is written by two experts in the field and I'm sure you will enjoy reading it!

Brisbane   Wil van der Aalst
January 2015

Preface

Nowadays, software and systems engineers are facing many challenges when they develop new systems. Society is expecting those systems to have high quality, provide exciting functionalities, and be produced at low cost. To cope with the growing complexity and diversity of engineering problems, the adoption of systematic and disciplined approaches to deal with requirements is of paramount importance. Over the past years, researchers and practitioners, most notably from the software and information systems domain, have contributed to create an extensive body of knowledge related to the engineering and management of requirements.

Although requirements engineering has essentially grown in the software and information systems domain, this book aims to have a broader perspective, since the process of characterising, for instance, a building, automobile, boat, or house includes many similar issues to those found to construct a software-intensive system.

Many of the methods and techniques discussed in this book are not exclusive for the software and information systems domain and can thus be applied in any engineering project, whatever branch or field. Project stakeholders are increasingly realising that requirements must be correctly handled if the project is to be a success. However, there are still many engineering projects where requirements are not adequately addressed, since the teams are not prepared for the task, due to lack of training, coaching, or experience.

This book focuses on various topics related to engineering and management of requirements, in particular elicitation, negotiation, prioritisation, and documentation (whether with natural languages or with graphical models). The book provides the reader with methods and techniques that help to characterise, in a systematic manner, the requirements of the intended engineering system. It was written with the goal of being adopted as the main text for courses on requirements engineering, or as a strong reference to the topics of requirements in courses with a broader scope. It can also be used in vocational courses, for professionals interested in the software and information systems domain.

We hope that readers will find this book valuable, either for understanding the rationale behind the methods and techniques, or as a handbook to support the approaches to be adopted in their next engineering projects. Visit the "Requirements in engineering projects" companion website at www.springer. com/RequirementsEP to find valuable educational material for students, lecturers, and researchers. To contact us, the authors, please send an email to RequirementsEP@springer.com.

Braga   João M. Fernandes
Guimarães   Ricardo J. Machado
February 2015

Acknowledgments

We thank our institution, Universidade do Minho (www.uminho.pt, Portugal), for providing us with many of the resources that were used to go ahead with this book project. First, the knowledge and experience that we have in requirements engineering is the result of the educational and research initiatives that we developed while professors at UMinho. Second, UMinho funded part of João's sabbatical year during 2013–2014, which was used to write a large part of the book. Finally, access to the UMinho bibliographic resources was fundamental for organising most of the contents of the book.

Part of this book was written in Florianópolis, Brazil, while one of us (again João) visited UFSC for a period of 5 months. We would like to thank UFSC for having provided extremely favourable working conditions, which allowed João the tranquility and the appropriate context in which to dedicate himself to the writing of this book. Access to the bibliographic resources at UFSC was another element that proved crucial to support the process of writing. This experience was also instrumental to decide to try to publish a version of this book in the Portuguese-speaking market, which obviously includes Brazil.

We would like also to thank all the students who took part in the courses mainly dedicated to the requirements engineering subjects that we have run since 2000 at UMinho, namely Requirements Engineering and Management, Analysis and Design of Information Systems, and Software Engineering for Information Systems, part of the Master degrees on Informatics Engineering, Information Systems Engineering, Electronics and Computer Engineering, and Informatics and Telecommunications Engineering. We acknowledge also the software engineers that attended the requirements engineering and management training course delivered by us in several occasions (Fernandes et al. 2009). Much of the contents presented in this book evolved during the various instantiations of those courses, allowing them to be validated in a real educational environment (Machado et al. 2014).

We acknowledge all the industrial and scientific partners with whom we have been able to experiment and assess several requirements methods and techniques. We particularly acknowledge Bosch Car Multimedia Portugal (former BARP—

Blaupunkt Auto-Rádios Portugal) for having supported a long-term research cooperation agreement with us, since the first UMinho R&D project with BARP in 1999. Within HMIExcel,[1] a joint Bosch/UMinho project, a considerable research effort on requirements engineering has been carried on. The new doctoral programme in Advanced Engineering Systems for Industry (AESI), recently set up at UMinho and supported by Bosch Car Multimedia Portugal, adopts a multidisciplinary systems engineering approach based on a strong and exhaustive training in requirements methods and techniques. AESI is specially targeted for preparing Ph. D. students for a career in industry and was the major Portuguese doctoral programme in cooperation with the national industry financed by the Portuguese Science Foundation (FCT) in 2014.

We would like to acknowledge the excellent service offered by the website linguee.com that provides a great variety of contextual translation examples. This tool proved extremely useful since we are not native English speakers.

The illustrations that make this book much more beautiful were made with great enthusiasm and dedication by Mônica Lopes, a Brazilian artist. We have to thank her for all the commitment that materialised in many email messages and Skype calls, to achieve a result that combines the quality of the drawings with the messages they transmit in the context of the book. We are sure that the reader will appreciate Mônica's excellent drawings that help to capture in a fun and challenging way many of the aspects related to the book's topics.

This book benefited from the excellent support provided by the Springer staff. The whole process associated with the book was extremely fast and professional. Nathalie Jacobs was very fast to analyse and accept our book proposal. Anneke Pot, our editorial assistant, was always very diligent to respond to all our questions and doubts related to the production of the book. It has been a pleasure to work with them both. We are grateful also to Edwin Beschler for proofreading the book and offering numerous useful suggestions and corrections.

---

[1]This book was partially supported by the HMIExcel project—a joint Bosch/UMinho research and technological development initiative aiming at critical R&D in the framework of the development and production cycle of advanced multimedia systems for the automobile industry.

Contents

Acronyms

AHP	Analytical hierarchical process
AIDS	Acquired immune deficiency syndrome
CSCW	Computer-supported cooperative/collaborative work
DFD	Data flow diagram
DNA	Deoxyribonucleic acid
ERP	Enterprise resource planning
FEM	Finite element analysis
GNU	GNU's not Unix
JAD	Joint application design
JPEG	Joint Photographic Experts Group
KA	knowledge area
Laser	Light amplification by stimulated emission of radiation
LED	Light emitting diode
MP3	Moving Picture Experts Group Layer-3 Audio
NBA	National Basketball Association
NBC	British Broadcasting Corporation
PDF	Portable document format
Radar	Radio detection and ranging
RAM	Random-access memory
RFID	Radio-frequency identification
SMTP	Simple mail transfer protocol
Sonar	Sound navigation and ranging
SWEBOK	Software engineering body of knowledge
TNT	Trinitrotoluene
TV	Television
UML	Unified Modeling Language
UNICEF	United Nations Children's Fund
USA	United States of America

Chapter 1
Presentation of the Book

Abstract This chapter makes an overall presentation of the book. In addition to a first introduction to the problems associated with the requirements, the objectives of the book and how it is structured are presented. There is also a discussion about taxonomical issues, in which the terms most commonly used in this book are introduced and defined. The chapter concludes with a short biographical note about the authors.

1.1 Introduction

Engineering is a discipline that is intimately related to the concept of project. It is throughout a project that the engineer applies in a more visible way his technical and scientific knowledge, to solve the problems and to achieve the objectives which he is confronted with. The project is one of the most noble areas within the scope of engineering, regardless of the technical branch where it originates (for example, civil, chemical, electrical, mechanical, aerospace, industrial, railway, mechatronics, transport) or the nature of the economical activities where it occurs (for example, building construction, car industry, pharmaceutical industry, consumer electronics). Although the project constitutes a classical topic in the scope of engineering, the importance that here is assigned to the project is justified by the fact that it represents the context where the need to cope with the requirements occurs.

In order to solve a given problem with a project, the engineer needs to devise a new system (structure, machine, product, process or service) that meets the objectives and satisfies the requirements originally laid down. The project seeks to avoid failure, through the planned release and control of the activities and the management of the contingencies and unplanned issues that may endanger the objectives (Illustration 1.1). The project is also temporary, since it is intended to create something different from everything that exists or to change significantly something that already exists. The result of the project can then give rise to the production of multiple replicas of the system that was designed. The regular production of a system, outside the context of a project, is referred to as *current activity*.

Since the project is one of the common areas to the different engineering branches, it is expectable that the different scopes and objectives of a project and the technical

© Springer International Publishing Switzerland 2016
J.M. Fernandes and R.J. Machado, *Requirements in Engineering Projects*,
Lecture Notes in Management and Industrial Engineering,
DOI 10.1007/978-3-319-18597-2_1

Illustration 1.1 The project is a noble area of engineering and it involves the definition of the activities in a given time horizon, allowing a specific system to be developed

and organisational specificities of the application domains (nature of the economical activities where it is conducted) promote several possible definitions for the project concept. However, there are some common aspects that seem to gather some consensus. Succinctly, a **project**, within the scope of engineering, is a temporary enterprise, composed of various coordinated activities with well-defined start and stop dates, undertaken to create a unique system. Generally, issues like time, cost, performance and quality are extremely important. Additionally, similarly to what happens with its activities, the project is also developed by people, in accordance with a given purpose, executed by following a plan, and controlled based on defined criteria, with the objective of respecting the restrictions that are relevant. A project is different from the current activities in three primary aspects:

1. contrarily to the current activities, each project produces systems with **unique characteristics** and that differentiate them from similar systems;
2. each project is only developed between an initial moment and a final moment that, when achieved, definitely **closes the project**, something that may not occur for the activities;
3. the organisational and structural means, created during the project, normally **disappear or are modified**, when the project ends.

"How does a project get to be a year late? One day at a time."
Frederick P. Brooks Jr. (1931–), computer architect

To correctly understand the principal motivations behind this book, it is important to clarify the meaning that is associated with the term 'systems engineering'. This clarification requires a reflection about the meaning of fundamental and related concepts like system and engineering.

A **system** can be defined as an identifiable and coherent set of components that interact to achieve a given objective. It is a collection of elements, in which each one is interrelated to at least another one and that possesses distinct properties from the properties of its parts when considered individually. The saying "the whole is greater than the sum of its parts" defines implicitly the existence of systems, as long as they possess at least one emergent property, that is, a property that depends on the structure of the system, emerging from the way the different parts interact among them to constitute that system.

The conceptualisation of a system implies the existence of a frontier that separates the system from the rest of the world. It is through that frontier that the system interacts with the surrounding environment, resorting to a physical or simply logical interface. This separation of the system from its environment reflects a relativist vision of the world, since its conceptualisation depends on the way the engineer wants to comprehend it and the utility that she intends to assign it. This less classical, but simultaneously more useful, view of a system, justifies, from the perspective of the activities of engineering projects, a set of restrictions for the concept of system (Thomé 1993a).

Despite the fact that systems exist in the world, for engineers this finding is frequently inconsequent. In engineering, the critical issue is the definition of the frontier that separates the system from its environment, taking into account not only the emergent properties of the system, but fundamentally the objectives that the engineer has in her mind.

Any set of parts can be seen as a system, which implies that the system is in the eye of the beholder according to the intended purpose. In engineering, the system is a technical human-made entity that is somehow organised to serve a given purpose. In forward engineering, it is common to use the term 'system under development' to designate the system that needs to be developed.

The purpose of the system is only achieved by the system as a whole. This means that, if this was not the case, the engineer, with his capacity to model the world according to what is more convenient from him, would have formed it in a different way. This represents an alternative perspective to understand how the engineer positions himself when faced with the project to devise a given system.

The existence of systems is ephemeral, in the sense that they exist, with the initial emergent properties, solely during a short period of time. This degradation of the systems results from three distinct factors: (i) according to the laws of the Nature, any system has an end, due to the material renovation cycle; (ii) due to the relativist perspective of the engineer, a set of interrelated elements that, in a given instant, are seen as a system may, in a subsequent instant, no longer be viewed as such; (iii) by the pressure of the surrounding environment, the systems are "forced" to evolve during their lifecycle to adapt to new requirements. These three factors justify the continuous concern of the engineers over the project, to guarantee a solution that is able to cope with changes, allowing the system to survive in an environment that is constantly changing.

The relativity in the definition of the frontier between the system and its environment, and also the fact that knowing the purpose of the system is a demand of the project that the engineer must cope with, allows one to see the systems as hierarchies of: (i) macro-systems, when one considers, for instance, that the system and its environment both form another system; (ii) subsystems, when the elements of the system under consideration are themselves seen as other systems. However, it is important to consider that the purpose of the macro-system is not necessarily the same as the systems that it is composed of.

In this book, the notion of system refers to the result of the execution of a project. Besides this notion, the engineering activity requires dealing with another notion of system; the one that considers the engineering process as a system. Thus, it becomes possible to define **engineering** as being the application of a systematic, disciplined and quantifiable approach to the analysis, design, implementation and exploitation of systems, resorting to knowledge, principles, techniques and methods that originate from the empirical-scientific advances, in an ethical context to satisfy the necessities of the human development.

Engineering consists in applying scientific and technical knowledge to solve problems. Engineers are responsible for conceiving and constructing useful artefacts that permit to solve the problems which they are confronted with as professionals. Engineering aims to fulfil the necessities and expectations of the society in general and to solve its problems, with the perspective of reducing costs, increasing the reliability or performance of a system or process, and even to satisfy consumers, as a result, for example, of changes in their daily routines.

Engineering is targeted to results, but in scenarios with time and financial restrictions. These restrictions, which all engineers are subject to, imply that they cannot have a perfectionist approach. Engineers must assume certain trade-offs in order to find a solution that satisfies the problem, but that obeys to the identified restrictions. Hence, engineering is the application of scientific and technical knowledge in the analysis, design, and construction of systems with the aim of satisfying the necessities of the social development.

"Because startups often accidentally build something nobody wants, it doesn't matter much if they do it on time and on budget."

Eric Ries (1979–), entrepreneur

The *systematic approach* takes the application of the notion of system to the limit, allowing one to conceive of anything as a system. It represents one of the flags of the modern engineering, namely because it considers many aspects that are fundamental for complex systems: abstraction, reductionism/holism, complexity, and flexibility.

The complexity of a system depends not only on the number of its parts, but essentially on the way they interact among themselves. This fact suggests that the engineer should explicitly control the dimension and the heterogeneity, since in this way she helps make it possible for the system to be effectively handled as a multi-faceted whole, allowing its emergent properties to be naturally expressed.

Being aware of the abstraction level in which the project is developed is crucial to control its complexity. In fact, to effectively control the inherent complexity of the system, it is important to decide which abstraction level is the most adequate at each moment. As a rule, one must resort to higher abstraction levels, so that, when hiding the details, one understands more easily the system as a whole.

When managing the complexity and deciding about the abstraction level to adopt, one must always weigh the advantages and disadvantages when choosing one of two possible orthogonal views:

1. **Reductionist**, when the system is decomposed into smaller parts, in order to control the complexity and to concentrate the project effort on multiple subsystems, but each one simpler. One must be aware that the emergent properties may disappear when the system is decomposed.
2. **Holistic**, when, to avoid handling details that may disperse the attention towards the emergent properties, one considers the system as a whole. This vision must be complemented with a redefinition of the system frontiers, which in practical terms contribute to relativising the duality system/subsystem.

Some flexibility should exist when choosing the process model to be adopted, considering various project alternatives and the possible scenarios of intervention (normality and exceptionality). This flexibility must result from the application of the feedback mechanisms and the iteration of the process model itself. Emergence of the systematic approach has not only influenced the methodological approach of engineering itself as a whole, but it has also provided the foundations to the appearance of a new engineering field now known as systems engineering. The concept of systems engineering can be decomposed according to two perspectives:

1. **Engineering of systems**, understood as the engineering of the systems whose parts imply more than one technology (software, hardware, mechanical, etc.). According to this interpretation, any engineering of non-homogeneous systems can be seen as an engineering of systems. This definition of systems engineering is product-oriented.
2. **Systematic approach to engineering**, seen as a disciplined and methodic approach to engineering, regardless of the technology of its components. Thus, any engineering can be seen as a systems engineering, as long as it follows the principles of the systematic approach. This definition of systems engineering is process-oriented.

Actually, systems engineering combines the two orthogonal perspectives, that is, systems engineering consists in applying the systematic approach to the engineering of systems. Its domain is the engineering of solutions for problems associated with complex and heterogeneous systems. Thus, **systems engineering** can be defined as the interdisciplinary branch of engineering that is devoted to developing and managing complex and heterogeneous systems, according to a systematic approach.

As engineering systems are becoming more and more complex, it is no longer possible to assign to one person the responsibility of developing them. The project of a complex system must now be done in teams, which implies the need to detail information about their characteristics and to provide mechanisms that will make communication effective, either among the engineers implicated in the project or between them and the stakeholders (namely the users).

However, understanding the requirements of a system is among the most difficult tasks to be accomplished with success in the scope of systems engineering. Unfortunately, there is no unique and stereotyped approach to execute that task and it is not rare that both students and engineering professionals consider that handling requirements does not require special competencies.

Many of the methods discussed in this book are generic, so they can be applied to whichever engineering branch, field, or speciality in which the project has been conceived. However, in this book, most examples and many concepts related to engineering systems are specific to the software and information systems domain, due to the predominant role that it has in complex systems (for instance, airports, hospitals, intelligent buildings, automobiles, banking systems, transports). Software is, in a large number of systems, the element that determines the respective value. Some authors, like Kossiakoff and Sweet (2013, p. 361), indicate that the development of software systems, due to their abstract nature, is different and more complicated than the development of systems composed mainly of hardware parts.

1.2 Objectives of the Book

This book focuses on various topics related to requirements engineering, in particular elicitation, negotiation, prioritisation, and documentation (whether with natural language or with graphical models). The book provides the reader methods and techniques that help to characterise, in a systematic manner, the requirements of the intended system.

> "There is no book so bad that it does not have something good in it."
> *Miguel de Cervantes (1547?–1616), novelist*

This book was written with the primary objective of being adopted as the main text in undergraduate and graduate courses on requirements engineering, or as a strong reference to the topics of requirements in courses with a broader scope (e.g., systems engineering, software engineering, information systems engineering, project management). This book can also be used in vocational courses or recycling training sessions, for professionals in the business of software and information systems. Hence, those professionals have the opportunity to address some thematic of their interest in a systematic and organised manner.

This book covers topics related to requirements of engineering systems and was written to ensure a reasonable coverage of the issues that should be addressed in a university course. This main goal has determined that the book does not include all topics related to the requirements in engineering projects. This book aims to provide the reader the essential topics. Obviously, one is not saying that the issues dealt with in less detail or even ignored in this book are not relevant, but only that they were placed in a secondary plan in accordance with the main goal. This way, this book is more complete in the topics related to requirements than the books that include only one or two chapters for requirements. This is the case in the field of software engineering, with some books, like for example: (Ghezzi et al. 1991; Sommerville 2010; Pressman 2009; Pfleeger and Atlee 2009). On the other hand, the books (van Lamsweerde 2009; Pohl 2010) are very dense and complete and include almost everything related to requirements. However, they cover too many topics, some of which have a reduced applicability in many real projects.

This book includes, at the end of each chapter, a recommended reading section, to provide the reader pointers to extra material on the various topics addressed. Additionally, the many references made throughout the text, which in the case of books include whenever possible the respective pages, allow the reader to access more literature that helps certainly to complement the contents of this book with other visions. Each chapter includes also a set of exercises that the reader can explore in order to test her knowledge on the addressed topics. Many of those exercises are inspired in the material presented by Naveda and Seidman (2006), which includes interesting exercises for all KAs of the 2004 edition of the SWEBOK (Abran et al. 2004) (more about SWEBOK in Sect. 2.2).

When the reader finishes reading this book, it is hoped that he will be able to:

- establish and plan a requirements engineering process within the development of complex engineering systems;
- define and identify the types of relevant requirements in engineering projects;
- choose and apply the most appropriate techniques to elicit the requirements of a given system;
- conduct and manage negotiation and prioritisation processes for the requirements of a given engineering system;
- document the requirements of the system under development, either in natural language or with graphical and formal models.

1.3 Structure of the Book

This book is structured in eight chapters. Chapter 2 introduces the software and information systems domain, addressing its most important subjects, according to two orthogonal perspectives adopted in systems engineering (product and process). Chapter 3 discuss some definitions for the 'requirement' concept and presents several types of requirements that one should consider in engineering projects. In Chap. 4, requirements engineering is presented and characterised, namely the activities that its process is composed of. Chapters 5–8 describe in detail the three activities that are considered to be fundamental for the requirements engineering process: elicitation, negotiation and documentation. Chapter 5 identifies and characterises some of the stakeholders that may be implicated during the development of a system and presents many techniques and approaches for eliciting requirements. Chapter 6 discusses how to negotiate the requirements amongst the stakeholders of a system to collectively decide those that must be implemented. One also addresses the importance of assigning priorities to requirements. In Chap. 7, various practical recommendations for writing requirements in English are suggested and a structure for documenting requirements is analysed. Phenomena related to ambiguity in natural languages are also debated, with the aim of emphasising the attention that must be paid when using a language of that nature for documenting requirements. Finally, Chap. 8 addresses topics related to requirements documentation, resorting to modelling approaches that are typical in the software and information systems domain, with a natural focus on the most relevant models for the activities associated with the requirements engineering process.

1.4 Taxonomical Issues

This book avoids the unrelieved use of 'he'. It also avoids the 'he or she' and the distracting 'he/she' that appear in many books. The adopted solution is to use 'he' and 'she' (or similarly 'him' and 'her', or 'his' and 'her') randomly, as the text goes along. When, for example, the term 'engineer' is being used one is referring to either a man or a woman that possesses the habilitations and the skills necessary to develop engineering acts. The illustrations in the book include a female requirements engineer, but this was also a choice to make clear that engineering is not a male discipline. Indeed, engineering traditionally attracts more men than women, but we firmly believe that this situation needs to be more balanced, by engaging more female students in this discipline.

> "Excellence is the best deterrent to racism or sexism."
> *Oprah G. Winfrey (1954–), talk show host*

One of the inevitable consequences of the fast evolution of the software and information systems domain lies in the fact that the same word is frequently used by different persons with diverse meanings. It is common to see an engineer asking another one what is the real meaning of terms like 'model', 'design', 'component', 'process', or 'method'. The list of terms that fit here is certainly long. Additionally, as happens in many other knowledge areas, common words gain specific meanings when used technically. To alleviate the negative impacts of this question, definitions for various terms that are regularly used in this book are next presented. The principal terms used in this book, as well as the acronyms, are presented in the glossary. Some terms were analysed with especial caution and so the choices made are now explained.

Systems go through various phases throughout their existence. A **phase** is an abstraction, in the time domain, of a set of activities, that is, it is an useful concept for aggregating activities from a practical point of view (Whytock 1993). In reality, a clear division among phases may not exist, since it is possible that the activities are executed in a cyclic, iterative, or incremental way. Additionally, even the designations of the most typical phases are not consensual. There are several alternatives to name them, which sometimes creates some misunderstandings.

The system **lifecycle** is the sequence of phases that starts when the system is mentally conceptualised, that is, as soon as the idea, necessity, or will to construct it appears, and that ends when it is decommissioned. The lifecycle is associated with the set of valid acts, realised, over the useful life of the system, with the objective of idealising, developing, and using it.

Systems are developed to solve a given problem. Hence, the **development** refers to the lifecycle phases responsible for the construction of the system, including analysis, design, and implementation. The tasks that precede the development, such as for example feasibility studies, as well as those that happen after the system is constructed, such as maintenance and the effective operation of the system, are excluded.

According to (Meyer 2013, p. 703), **maintenance** of software systems covers all development activities occurring after the first release of an operational version. It includes the adaptation to new platforms and environments, correction of defects, addition of new functionalities, removal of unless functionality, and quality improvement.

Development can be followed according to different approaches, since some of them more are structured and planned and others more iterative and incremental. In general, a **process** is a sequence of activities executed with the objective of achieving a given result (for example, manufacturing process or chemical process). For instance, the industrial process in a factory corresponds to the set of activities executed to produce a given consumer good. Processes involve methods, tools, and persons.

"If you can't describe what you are doing as a process, you don't know what you're doing."

W. Edwards Deming (1900–1993), electrical engineer

In the software and information systems domain, a **software process** (or development process) is a sequence of steps needed to develop or maintain software (Humphrey 1995, p. 4). The result of the software process is materialised through several artefacts. An **artefact** is a tangible entity used or produced during the development process. Some artefacts (e.g., models) document the functions of the system or part of it, while others are related to the development process itself, such as, plans, intermediate reports, or quality reports. It is possible to consider that the system that results from the development process is itself an artefact.

A **process model** is an abstraction of a process, describing it in accordance with a given modelling perspective. A process model organises, sorts, and relates the way the distinct phases and tasks must be pursued during a given process. The principal function of a process model is to determine the order of the phases during the systems development and to establish the transition criteria to progress from on phase to the next one (Boehm 1988).

It is possible to divide the lifecycle of a system in the following generic phases[1]:

- **feasibility study**, whose aim is to decide about the relevance and the viability of initiating the development of a new system, taking into account, namely, economic, technological, and commercial factors;
- **analysis**, where the system functionalities and all decisions that may restrict the design and the implementation activities are documented;
- **design**, in which the internal structure of the system is defined and one assigns to each of its components the functionality to be supported, in order to fulfil the specifications provided by the analysis phase;
- **implementation**, where the system is constructed according to the design directives and decisions;
- **testing**, whose objective is to evaluate if the system fulfils what is expected and to identify defects that the system may possess;
- **utilisation**, in which the system is operating in the real environment;
- **maintenance**, whose goal is to correct the detected defects and to enhance the system, so that it continues to be useful, after it starts to be used.

Figure 1.1 aims to relate and show the scope of some of the most important terms defined in this section. Please note that the feasibility studies may require efforts that are typically framed in the analysis phase. Analogously, maintenance may require the execution of some development activities (analysis, design, implementation, testing), if new requirements need to be incorporated in the system. Actually, the evolution of a system assumes a constant alternation between (re)development (i.e., maintenance) and utilisation.

A **method** is a set of generic guidelines that conduct the execution of a given set of activities, within the scope of a specific lifecycle phase. There are methods specifically devised for the analysis phase (Coad and Yourdon 1991), the design phase

---

[1]Although no specific process model is assumed, one can suppose a process that follows, for instance, the waterfall model (see Sect. 2.4.1), where there is a more explicit division among the phases when compared with other models.

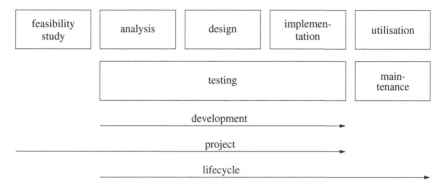

Fig. 1.1 Lifecycle phases of a system

(Budgen 1994; Albin 2003) and even the testing phase (Kit 1995; Perry 2006). A method is composed of a series of steps and represents a systematic way of conducting with a well-defined phase of development a computer-based system (Morris et al. 1996, p. 22).

> "Art and science have their meeting point in method."
> *Edward Bulwer-Lytton (1803–1873), novelist*

The term 'method' is sometimes used, in the specific case of software engineering, as a synonym of function, routine, procedure, or service. In many object-oriented computer programming languages, a method is a function that a given object (or class of objects) is apt to execute. However, the distinction of the precise meaning is easily obtained based on the context where the term is used.

In this book, someone engaged in the project of a system, with technical roles, is designated as systems engineer or developer. In the context of the execution of a system project, it is possible to identify specific technical roles associated with responsibilities restricted to the project phases. An **analyst** (or requirements engineer) is a developer specialised in the tasks related to the analysis phase. In particular cases, where one needs a more clear delimitation of the acting scope of the developer, terms like 'architect' (for design activities), 'programmer' (for implementation activities), and 'tester' (for testing activities) may also be used.

1.5 About the Authors

João M. Fernandes is Full Professor at Dept. Informatics/ALGORITMI Research Centre, Universidade do Minho, Portugal. He conducts his research activities in software engineering, with a special interest in software modelling, requirements engineering, and embedded software. As part of his research and teaching activities, his work is focused on the methodological and technologic aspects related with the use of a multi-perspective, model-driven approach for developing software systems. Within his research and teaching activities, he maintains regular collaborations with the industry.

He was invited professor/researcher at U. Bristol (UK, 1991), Åbo Akademi (Turku, Finland, 2002–03), ISCTEM (Maputo, Mozambique, 2003), U. Algarve (Faro, Portugal, 2004–06), U. Århus (Denmark; 2006–07), and UFSC (Florianópolis, Brazil; 2013).

He has authored more than 100 scientific publications published in journals, books and conference proceedings and is co-editor of the book "Behavioral modeling for embedded systems and technologies: applications for design and implementation" (IGI Global, 2010). João is member of the Editorial Review Board of the Journal of Information Technology Research, IGI Publishing, since Jun/2007. He has been involved in the organisation of various international events, including ACSD 2003, DIPES 2006, GTTSE 2009, PETRI NETS 2010, ACSD 2010, the MOMPES workshops series (2003–2012), and ICSOB 2015. He is a regular reviewer for scientific journals and conferences. He also regularly serves as a member of the program committees of international conferences and workshops.

For more information, please access www4.di.uminho.pt/jmf.

"The writer writes only half the book; the other half is with the reader."
Joseph Conrad (1857–1924), novelist

Ricardo J. Machado is Full Professor at Dept. Information Systems and director of the ALGORITMI Research Centre, Universidade do Minho, Portugal. He conducts research activities and has coordinated more than 50 research projects in the field of information systems engineering, with a special interest in modelling approaches for analysis and design, and in process and project management life-cycles, resulting in more than 150 publications. Within his research and teaching activities, he maintains regular collaborations with the industry, currently acting as president of the General Assembly of TICE.pt (the ICT Industrial Pole formally recognised by the Portuguese Government, as part of the QREN's Strategies for Collective Efficiency).

He is the founding coordinator of the SEMAG Research Group (2003–) at the ALGORITMI Research Centre and the founding coordinator of EPMQ Laboratory (2008–) at the CCG/ZGDV Institute. He coordinated the participation of the University of Minho within the Carnegie Mellon—Portugal Programme (2006–12).

He acted as president of the CT128 Technical Committee (2004–11) at the Portuguese Institute for Quality, responsible for analysing the documents produced by

Illustration 1.2 João is passionate about cycling, a non-polluting form of transportation that he uses regularly to move in urban contexts. Ricardo likes to take pictures for the pleasure of always perceiving that everything around us looks different when one starts to see the world as a photographer

ISO/IEC JTC1/SC7 and by CEN/CENELEC TC311. He was also general coordinator for the Region 8 (Europe, Middle East, and Africa) of the IEEE Computer Society (2007–08), and also coordinator of the Portuguese representation in the IFIP TC10 Technical Committee (2006–09).

He is a founding member of the IFIP WG10.2 Working Group on Embedded Systems and of the IEEE-IES Technical Committee on Education in Engineering and Industrial Technologies. He is a Senior Member of IEEE Computer Society and of the Portuguese Engineers Association—Informatics Engineering Society. He received the 2009 IEEE MGA Achievement Award.

For more information, please access www3.dsi.uminho.pt/rmac.

João and Ricardo have other interests beyond the engineering field, as Illustration 1.2 shows.

Chapter 2
Software Engineering

Abstract Software engineering is an engineering discipline that is focused on all aspects concerning the development of software-based systems. This chapter begins with an explanation of the contributions of software engineering to the issues related to requirements, discussing the possibility of adopting their methods on projects of other engineering disciplines. The chapter also characterises the software engineering, identifying and describing the fifteen knowledge areas of the SWEBOK guide. Additionally, the most relevant characteristics associated with the software are discussed. Finally, some of the most popular development process models are presented.

2.1 Contributions for Requirements Engineering

The advances and progresses that were made during the last 40 years in the software and information systems domain are, whatever the perspective, extraordinary. Despite some negative propaganda (for instance, the eternal software crisis or the non-event that was the *year 2000 bug* (Y2K bug), there is a huge number of success cases that changed human daily habits, in a significant or even drastic way. Nowadays, software is present not only in traditional personal computers or in highly-sophisticated supercomputers, used for scientific purposes or for executing governmental activities, but also in mobile devices, like cellular phones or tablets, or in devices responsible for routing in computer networks.

The existence of a knowledge area related to software engineering is due to the growing complexity of the developed systems and to economic factors that prompt software producers to try to optimise the processes. Therefore, they can gain a competitive advantage to be better positioned in the markets where they operate. All these issues have influenced the evolution of this area, which started to take shape at the end of the 1960 decade. There are of course problems associated with software development and many software projects are not ready on time and cost much more than was initially planned. It is due to the fact that those problems can occur that one needs to adopt an engineering approach.

© Springer International Publishing Switzerland 2016   15
J.M. Fernandes and R.J. Machado, *Requirements in Engineering Projects*,
Lecture Notes in Management and Industrial Engineering,
DOI 10.1007/978-3-319-18597-2_2

Software engineering is a discipline composed of theories, methods, approaches, and tools needed to construct software systems. Software engineering, like the other engineering branches, needs to base its activity on *quality*. This implies that everything that is related to engineering, including obviously the developed systems, must possess quality, in order to satisfy the client. In a simple way, quality is the compliance with the requirements. The main objective of software engineering is to develop software systems with quality, fulfilling the budget, meeting the deadlines, and satisfying the real needs of clients by solving their problems.

> "The function of good software is to make the complex appear to be simple."
> *Grady Booch (1955–), software engineer*

To cope with the growing complexity and diversity of real-world problems and to changes in requirements during the development process, the construction of systems in the software and information systems domain needs to adopt engineering approaches, to improve the quality of those systems. This does not imply that the system is for sure optimal, but that normally it possesses a good quality and is produced in a controlled way and limited by the available budget.

In this context, over the last years, the software engineering discipline has accumulated an extensive scientific body of knowledge related to the requirements and their problems. This reality results from the need to control the enormous uncertainty and the high number of risks normally associated with this domain, due namely to the intangible nature of the information technologies. Operationally, the success of projects in the software and information systems domain requires an equilibrium between the capacity of reading the surrounding reality (environment) and the ability to socio-technically act upon it. These demands refer to the relationship with the stakeholders and to the technical decisions associated with the project management.

Based on this track taken by the software engineering discipline, the dimension and importance of the scientific community that is focused on the subject of requirements has grown tremendously. This community adopted the 'requirements engineering' concept, understood as the set of activities that in the context of developing a system allows one to elicit, negotiate, and to document the functionalities and restrictions of that system.

In fact, the process of characterising a building, automobile, boat, or house contains different aspects than those that are relevant in the process of defining the requirements of a system in the software and information systems domain. However, it also includes many similar issues. The methods and the techniques discussed in this book result from the contributions made by the requirements engineering scientific community. In most cases, those methods and techniques are not exclusive for the software and information systems domain and can thus be applied in any engineering project, whatever branch or field.

2.2 Characterisation of the Discipline

Software engineering, as a discipline, can be defined in different ways. In a simple way, one can say that it corresponds to the application of the engineering principles to the software development process. This line of reasoning was used by Fritz Bauer, when in 1968, he defined it as being the utilisation of the basic engineering principles to obtain, in a economically-viable way, reliable software that runs efficiently in real computers.

Software engineering is focused on all aspects related to the development and utilisation of software systems, from the initial steps of specification until maintenance. There are two software engineering aspects that deserve to be highlighted. On the one hand, engineers must be able to manage the development and industrialisation of useful and efficient system. On the other hand, software engineering is not just centred in the technical aspects associated with the development, but includes also activities related to managing the development process.

> "Leonardo Da Vinci combined art and science and aesthetics and engineering. That kind of unity is needed once again."
>
> *Ben Shneiderman (1947–), computer scientist*

One can also define software engineering as a computer science field that tackles the construction of software systems whose complexity requires a team of software engineers to build it (Ghezzi et al. 1991, p. 1). Here, the emphasis is put on the system complexity and, as happens in all engineering branches and fields, whenever the scale of the systems is changed, different problems and challenges arise. Some software systems have a long lifecycle, which explains the reason why they go through several versions in order to make sure that they adapt to new realities. The systems developed within the scope of software engineering are also used by different types of users and may include gigantic volumes of information.

Small software programs, written and maintained by a single person (the so-called heroic programmer), are not generally considered sufficiently complex to demand the use of an engineering approach. It is relevant here to distinguish computer programming from software engineering, since unfortunately there are still many persons, some professionally connected to computing, that consider both activities to be the same. A *programmer* writes complete software programs. It is very difficult, even literally impossible, for a single programmer to dominate all the facets of a software system, since its complexity largely exceeds the human mind capacities (Booch et al. 2007, p. 8). It is very likely, according to Laplante (2007, p. 4), that a software engineer spends less than 10% of his time in programming activities, using the remaining time for the other activities of the software engineering process. Regardless of the number being totally aligned with the reality, the relevant thing to understand here is its order of magnitude.

Illustration 2.1 A typical software engineering environment, with programmers, software architects, analysts, testers, and clients

The differentiation between software engineers and programmers is intimately related to the professional liability and accountability for acts of public trust in technologic interventions. It has been particularly difficult to convince the society and the professionals themselves that, in contexts of great complexity, the engagement of software engineers in the implementation phase is not justifiable to execute programming tasks. Software engineers should instead technically coordinate the work of the programmers. Metaphorically, software engineering is related to programming in the same way as civil engineering is associated with civil construction.

Software engineering includes also a specific management component that does not exist in programming. In a small project, with one or two programmers, the relevant aspects have essentially a technologic nature. In a longer project with a team composed of many members, it is indispensable to conduct management efforts, to plan and control the activities of the various professionals with differentiating roles, such as analysts, architects, programmers, and testers (Illustration 2.1).

In this book, **software engineering** is defined as the application of a systematic, disciplined and quantifiable approach in the context of the planning, development and exploration of software systems, that is, it is the application of engineering to the software domain. It is under this definition that appears the SWEBOK (software engineering body of knowledge) guide (Bourque et al. 1999; Abran et al. 2004), as an important reference to characterise the software engineering discipline. This guide is the result of an initiative jointly promoted by the IEEE Computer Society

Table 2.1 KAs of the SWEBOK

1	Software requirements	9	Software engineering models and methods
2	Software design	10	Software quality
3	Software construction	11	Software engineering professional practice
4	Software testing	12	Software engineering economics
5	Software maintenance	13	Computing foundations
6	Software configuration management	14	Mathematics foundations
7	Software engineering management	15	Engineering foundations
8	Software engineering process		

(IEEE-CS) and the Association for Computing Machinery (ACM). The SWEBOK was created to address the following objectives:

- to promote in the scientific community the existence of a coherent view of software engineering;
- to clarify how software engineering is related to other disciplines, such as computer science, computer engineering, project management, and mathematics;
- to characterise the scope and thematic contents of the software engineering discipline;
- to ease the access to the contents of the software engineering body of knowledge;
- to provide a reference for the definition of *curricula* and professional certifications in the software engineering discipline.

The SWEBOK guide structures the software engineering corpus according to the KAs shown in Table 2.1. KAs 11–15 were only introduced in third version of the guide (IEEE 2014).

Next, all the KAs are very briefly presented, to provide a generic idea of the subjects covered by software engineering. Thus, it is possible to relate requirements engineering (the main topic of this book) with the other activities carried out in the context of software engineering.

Most topics detailed in this book fall within the scope of KA1 (software requirements). This KA handles the elicitation, analysis, documentation, validation, and maintenance of software requirements. Requirements are considered as properties that the systems (still in project) may manifest later after development. The software requirements express the necessities and the restrictions that are put to a software system and that must be taken into account throughout its development. This KA is recognised as having primary importance for the software industry, due to the impact that its activities promote on stabilising and managing all the development process.

Software design (KA2) is the process where the architecture, components, interfaces, and other system (or its components) characteristics are defined. From the process perspective and within the scope of the software development lifecycle, software design is the activity in which the software requirements are handled with the purpose of producing a description of the internal structure and organisation of the system. From the product perspective, the final result of the process should describe the system architecture (i.e., how it can be decomposed and organised into components), the interfaces between the components, and the components with a level of detail that permits their construction.

Software construction (KA3) represents the fundamental act associated with software engineering. It consists in implementing software in accordance with the requirements and that works correctly, through a combination of coding, validation, and testing activities. The implementation of software is intimately related to design, since the former must transform into code the architectures conceived and described by the latter. This transformation tends to be more and more automatic, since there are tasks perfectly repetitive and mechanistic. Thus, it is in the software implementation phase that the utilisation of tools is more critical, to free the engineers from the less creative and error-prone activities.

Software testing (KA4) constitutes a mandatory part of the software development. Simultaneously, it is an activity in which one evaluates the software system quality and enhances it through the identification of the defects and potential problems. Testing includes, for instance, the dynamic verification of the software system behaviour in comparison with the expected one, using a finite set of test cases, especially chosen to cover the most critical situations.

Software maintenance (KA5) consists in introducing changes in the software system, after it was deployed and brought into operation, in order to (1) improve the system, (2) correct defects, and (3) adapt the system to a new context. The maintenance phase copes with the defects, the technological modifications, and the user requirements evolution. It is recommended to prepare maintenance in advance during the development phases, to ease the tasks that compose it. Although software maintenance is an area of software engineering, it has received a smaller attention by the scientific community when compared, for example, with design and construction.

Maintenance can be reactive, when the intervention is dictated by defects observed in the system, or proactive, whenever the intervention is performed before detecting the defects. In another dimension, maintenance can be oriented towards correction, which means that one attempts to detect and to repair the defects, or oriented towards improvement, with the objective of enhancing the system to accommodate new requirements or contexts of utilisation. The IEEE 14764 standard employs these two dimensions, as illustrated in Table 2.2, to divide maintenance into four categories: preventive, corrective, perfective, and adaptive.

> "Maintenance typically consumes about 40 to 80 % of software costs. Therefore, it is probably the most important life cycle phase of software."
>
> *Robert L. Glass (1932–), software engineer*

Change is inevitable when developing systems, due to new business conditions, modifications on the necessities of the clients and users, reorganisations of the development teams, and financial and time restrictions in the projects. Software configu-

Table 2.2 Maintenance categories

	Correction	Improvement
Proactive	Preventive	Perfective
Reactive	Corrective	Adaptive

ration management (KA6) aims to identify the configuration of the software system, in distinct moments of the lifecycle, to systematically control the changes of configuration and to maintain the integrity and traceability of the software system. This activity can be part of a more extensive process that aims to manage the software quality. The configuration management process of a given system over time can also be designated as version/release management. Software configuration management refers to the activities of control and monitoring that start at the same time as the project does and that terminate only when the system is no longer used.

Software engineering management (KA7) corresponds to software management activities (planning, coordination, measurement, monitoring, control, and communication), to guarantee that the software systems are engineered in a systematic, disciplined and measurable way. This KA is considerably distinct from the management practiced in engineering processes of other branches, due to the specificities of software and its process, like the intangible and abstract nature of the software artifacts and the very high rate of technological update that is required in the software industry.

The software engineering process (KA8) can be seen in two distinct perspectives. In the first one, the process is considered as a set of directives that guide how the professionals should organise and execute their activities over the project, for acquiring, developing and maintaining software systems. In the second perspective, one intends to evaluate and improve the software engineering process itself. Since the first perspective is already largely handled in the scope of other KAs, it is mainly within the second perspective that this KA contributes to. This explains why it is also designated, in a more restrictive way, but that simultaneously better characterises its focus, as engineering of the software process.

The use of software engineering models and methods (KA9) is fundamental to allow software systems to be engineered in a systematic, disciplined, quantifiable, and efficient way. Taking into consideration the abstract nature of software systems, the models constitute an indispensable tool when taking decisions in all the development process phases. The methods allow software models and other artefacts to be created and manipulated throughout the system lifecycle.

Quality must constitute a permanent concern of the engineers, since one expects engineering systems to possess high quality. The quality is related to the conformity of the system under development with the requirements. Software quality (KA10) is an activity that spreads all the software process and that requires the treatment of non-functional aspects like, for example, usability, efficiency, portability, reliability, testability, and reusability. The quality in the software must be seen as a transversal concern to all the software process.

"Quality is never an accident. It is always the result of intelligent effort."
John Ruskin (1819–1900), social thinker

The software engineering professional practice (KA11) has a highly multidisciplinary nature and is focused on topics related to professionalism, ethics, law, group dynamics, psychology, multiculturalism, communication, writing, presentation.

Software engineering economics (KA12) gathers contents of economic nature related to software systems in a business perspective. This KA includes topics like economics fundamentals (finance, accounting, inflation, time-value of money), life-cycle economics (portfolio, price, investment decisions), risk and uncertainty, and economic analysis methods (return on investment, cost-benefit analysis, break-even analysis).

KA13–KA15 are related to concepts and foundations of three disciplines that are critical to the success of the software engineer: computing, mathematics, and engineering.

2.3 Software

To understand the full scope associated with the software engineering discipline, it is convenient to figure out what is software, if viewed as an artefact or set of artefacts that result from the engineering process. In this section, the most relevant characteristics of software are presented.

2.3.1 Definition of Software

The term 'software' is relatively new and, according to Campbell-Kelly (1995), was used for the first time in 1959. Cusumano (2004, p. 91) says that Applied Data Research, a company founded in 1959, was the first one selling a software product separated from the hardware. At the beginning of the popularisation of computers (1950 decade), it was common practice to sell hardware and software as a unique system. Software was at that time viewed as part of the hardware and was designated as 'program code'. The emancipation of the software has its origin related to a relevant fact, from an historical point of view: the decision of american justice to demand IBM to distinguish, in its accounting documents, hardware and software, providing separate prices for each one (Buxmann et al. 2013, p. 4), something that has become reality since the 1970s.

Surprisingly, defining the concept associated with the 'software' term is not easy. A first attempt to define software can be made by a process of elimination. In a classic perspective, a computer consists of hardware and software and it only works in a useful way if there is a fruitful and symbiotic combination of those two parts. In computers that follow a classical architecture, the hardware, by itself, is not capable of realising useful tasks from the user point of view. In fact, it is necessary to provide some of the hardware components with a list of instructions (in the form of a program) that defines the task to be accomplished. Equally, software to be executed needs a hardware support. Galler (1962) proposes that everything that, in the users perspective, composes a computer, except the hardware, is the software. Galler's definition is elegant due to its simplicity and has additionally the advantage of putting the users as a central element. Additionally, it induces the need to define the concept of hardware, which is not especially difficult to formulate, due to its tangible nature. The **hardware** of a computer is composed of electronic and electromechanical components, including, namely, the processor, the memory, and the input/output devices. The hardware of a computer refers to the material components (integrated circuits, printed circuit boards, cables, power supplies, plugs, and connectors) and not to algorithms or instructions.

"Hardware and software are logically equivalent. Any operation performed by software can also be built directly into the hardware and any instruction executed by the hardware can also be simulated in software."

Brian Randell (1936–), computer scientist

A definition of software, elaborated without a process of elimination, is however necessary. According to Blundell (2008, p. 4), **software** refers generically to the *programs*, which include the instructions that are executed by the computer (more specifically the hardware), as well as the *data* that are operated by those instructions. For Ceruzzi (1998, p. 80), software is simply the set of instructions that direct a computer to do a specific task. Alternatively, Tanenbaum (2006, p. 8) defines software of a computer system as being the *algorithms* (instructions that describe in detail how to accomplish something) and their respective representations, namely the *programs*. A given program can be materialised in different physical means (punched card, magnetic tape, floppy disk, compact disc, etc.), but its essence resides in the set of instructions that constitute it and not in the support where it is stored. The software is the set of programs, procedures and rules (and occasionally documentation), related to the operation of a system that aims to acquire, store, and process the data. Software is the abstract element that, together with the hardware, constitutes the automatised part of a real-world system, implementing a stimulus-answer mechanism, with the objective of satisfying the needs of an entity external to the system.

An intermediate form between hardware and software is the so-called **firmware**, which consists of software embedded in electronic devices during their manufacturing. Firmware is used when it is expected, for example, that programs are never (or rarely) changed or when programs cannot fail in case of lack of power supply. Thus, firmware is typically stored in non-volatile memory. In some processors, the operation is controlled by a *microprogram*, which is a form of firmware.

It is important to understand that the most classic view about software, as being a program that executes in a personal computer, is nowadays far away from being the only one. Software is an integral and fundamental part of the so-called *embedded systems*, which are computer-based systems integrated in equipments (typically electromechanical). In an embedded system, the main actions are performed by the hardware, but software plays a major role. Embedded systems are developed for a specific purpose and the communication with the outside world occurs through sensors and actuators. Embedded systems normally run continuously and in real-time. The software that exists in those system is called **embedded software**, because it is integrated in a system that cannot be classified as being of software. For this reason, embedded software is not sold in an independent way (Kittlaus and Clough 2009, p. 6). The end users generally do not associate to this software type the same characteristics of the most traditional software. Instead, the users perceive the software as a set of functions that is provided by the system.

A modern automobile has today a significant percentage of its engineering effort related to the production of software. Pretschner et al. (2007) indicate that the BMW 7 series models implement 270 functionalities which the automobile users can interact with and that the software, in total, occupies around 65 Mb (65×10^6 bytes) of code in binary language (i.e., the program codified in the language that can be directly executed by the hardware). These authors predicted that, in 2010, the software of a top-of-the-range automobile could reach the 1 Gb (10^9 bytes) figure, but according to Ebert and Jones (2009) this fact occurred one year earlier. Cellular phones are nowadays equipped with much more software than the one that could be found some years ago in computers of large organisations or corporations. According to figures indicated by Ebert and Jones (2009), a top-quality mobile phone can posses 1,000,000,000 (10^9) lines of binary code, the same being true of aerial navigation systems or with software to control spacial missions. Even less sophisticated devices, like washing machines, low-end mobile phones, or *pacemakers*, have approximately 1,000,000 (10^6) lines of binary code. In factories, there are so many pieces of equipment and processes controlled by tailor-made software systems. The electrical energy that arrives at homes depends, in large measure, on software that controls its management and distribution. Due to the fact that the world is becoming more digital at various levels, the examples are almost endless, due to the omnipresence, sometimes unnoticed, of systems with software in the modern societies.

> "The future lies in designing and selling computers that people don't realise are computers at all."
>
> *Adam Osborne (1939–2003), computer engineer*

The first characteristic of a software system, that distinguishes it from other engineering systems, is its intangible nature. A software system is not a concrete and material entity, that is, it has no physical existence, contrarily to what happens with most systems from the other engineering branches (civil, mechanical, naval, chemistry, electrical). The software is not restricted by the materials properties, nor ruled by the laws of physics. Ghezzi et al. (1991, p. 17) indicate that software is soft, since it is relatively easy to change it. The softness of software, which is explicitly reflected in its name, results from its condition of intangible system. This characteristic has been, however, the cause of some of the problems associated with software development, since the changes imposed are often made without a careful analysis in terms of schedule, cost, quality, and impact.

Due to its intangible nature, software is developed, but not fabricated or constructed in the classical meaning of the term. In this book, a company that develops software and where many software engineers work is designated as a **software producer**. Often one uses 'software supplier', as a synonym, although this designation may have a more commercial connotation, since the one that supplies (i.e., sells, rents, lends, offers) something is not always the one that fabricates it.

Due to the intangible nature of software, in reality producing the first replica of a software system implies high costs, but the subsequent replicas are produced at much lower (in some cases, insignificant) costs. Additionally, copying software is an extremely easy operation, from a technical point of view, and that does not introduce loss of quality, since in the digital world one can consider that there are no differences between the original and the replicas. Due to these facts, the cost of software is essentially determined by the cost of the human resources necessary to develop it.

> "Software is like entropy: It is difficult to grasp, weighs nothing, and obeys the Second Law of Thermodynamics; i.e., it always increases."
>
> *Norman R. Augustine (1935–), aeronautical engineer*

A second characteristic of software is related to the fact that supposedly it does not wear out, in the sense that it does not lose its qualities over time. Although software does not wear out, in the physical sense of the term, it exhibits an enormous deterioration or degradation, derived essentially from the alterations that are introduced with the aim of maintaining its usefulness. Actually, the incorporation of new functionalities implies, almost inevitably, the introduction of defects in the software, which means that it eventually loses quality during its lifecycle.

2.3.2 Software Systems and Products

As already indicated, a system, in the context of engineering, is an identifiable and coherent set of components that cohesively interact to achieve a given objective. This definition permits that almost everything that exists in the universe can be seen as a system, which turns out to be true, since it is difficult to imagine something that could not potentially be viewed in a systemic perspective. Maybe an electron, if considered as an elemental particle (without components), could not be viewed as a system. However, for the engineer, what matters is not knowing whether something is a system or not, but instead if that thing is viewed by him as a system (Thomé 1993). This happens because the engineer has an interest in studying the properties of that entity as a system. It is the engineer that defines the frontier of the system with the environment, which makes the system definition not intrinsic to it, but rather dependent on the particular purposes and intentions of the engineer in each situation. As a consequence, the components that in a specific context constitute a given system may be just a subsystem of a wider system in a different context. The term 'system' is used here in a comprehensive way, including concepts like structure, machine, product, process or service. Terminological variants such as apparel, appliance, arte-fact, equipment, gadget, installation, instrument, object and organisation, may also be used for designing systems.

Here, two criteria are proposed to classify software systems (software-dominated or software-intensive systems): 'what is sold' and 'number of copies' (Xu and Brinkkemper 2007). Crossing these two criteria gives origin to the types of soft-ware systems shown in Fig. 2.1.

Whenever the final customer buys a given appliance that includes software, this is normally designated as *embedded software*. This term is used, either for a unique appliance (for instance, a satellite or a spaceship), or for devices produced in large numbers (for example, television sets or mobile phones). These devices correspond to heterogeneous systems (that include parts with different technologies), rarely being referred to as 'software products', even in the cases where the software parts correspond to the most important technological dimension of the system. Gadgets like digital cameras, smart phones, or printers are generically called 'consumer electronics products', in spite of including significant portions of software. The comparison factor is here the complexity of the technological effort. In the case of heterogeneous systems with software when viewed as products, the devaluation of the software technology results from the typical focus of the end users on the mechanical or electronic parts that exist in those tangible products, rather than on the software parts. Based on a mathematical analogy in relation to the product designation, in a heterogeneous system, the software corresponds to the neutral element and never to the absorbing one.

Whenever the software system uses exclusively software technologies (i.e., some-thing that can be designated as a pure homogeneous software system), then there are

Fig. 2.1 Classification of software systems

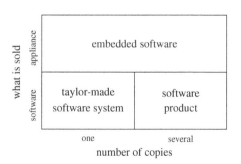

Illustration 2.2 A software product

two different types of systems. If the system is developed by request of a given client for satisfying his own necessities and expectations, then it is referred to as **tailor-made software system** (also called custom software system or bespoke software system). In this case, the principal objective is to satisfy the particular and specific needs of that client, without caring if it is equally useful for other clients.

If a software system is produced to be commercialised for, or made available to, the public in general, then it is designated as a software product, also called mass-market product. Generically, a **product** is a combination of (material and immaterial) goods and services that the supplier combines, in accordance with his commercial interests, to transfer established rights to the client (Kittlaus and Clough 2009, p. 6).

> "I already am a product."
>
> *Lady Gaga (1986–), singer*

According to this perspective, a **software product** refers to a homogeneous software product, being composed of the following three elements (Illustration 2.2);

- **programs**, whose instructions when executed offer the functionalities of the product;
- **data structures** that permit programs to access the necessary information for their execution;
- **documentation** that explains how to install, use, and maintain the programs.

Here, one defines a (computer) **program** as a text, written in a symbolic language capable of being interpreted by a computer, and composed of a set of operations that operate on the data and that obey the controls that stipulate the execution moments. This perspective of product is essentially technological, since it considers that the persons are not part of it, contrarily to what happens in the information systems with a socio-technical nature, in which are included the persons, as performers of parts of the organisational processes.

In business contexts, the development of a software product aims to obtain the highest possible number of customers, in the scope of the market segment that is identified for its commercialisation. The software producer has frequently the objective of selling massively, to maximise the respective market share and the economic incomes. Hence, a software product is developed in conformance with the common denominator of the necessities of the different users. If there is a tiny set of users that have a specific need, it is very likely that that need will not be included in the product. In any case, an underlying difficulty to develop a software product results precisely from not knowing in advance (i.e., during development time) who will use it.

> "Before new products can be sold successfully to the mass market, they have
> to be sold to early adopters."
>
> *Eric Ries (1979–), entrepreneur*

One can also classify software products in relation to the proximity that they have in relation to the hardware. Generically, two types of software products[1] can be considered: (1) *system software* responsible for managing the hardware resources of the computer; (2) the *software applications* that perform tasks that are useful for the end users.

A **system software** is composed of programs that interact in an intense and direct way with the computer hardware. Thus, it is normal that system software is not explicitly used by the end users. This type of software includes the operating systems, utilities to monitor the resources (to analyse, configure, and optimise the computer), *device drivers* and network software (web servers, email servers, network routers).

A **software application** (or software app) is a software product developed to support the realisation of the individual tasks of the persons and the execution of the organisational processes (government, industry, commerce, services). These applications to be executed use a computer (hardware) and a system software, for instance, an operating system. Therefore, one can say that, from the users point of view, they are (well) above the hardware level. These applications are essentially seen as productivity tools, that is, tools to enhance the human or organisational capa-

---

[1]Middleware could be considered as a third type, but in this book, a simpler solution was adopted. According to this first criterion, middleware is basically a generic designation used to refer to the software that executes between the system software and the software applications. The objective is to promote an easy development of distributed applications, since middleware serves to transfer informations and data among programs.

bilities. The software applications help people in various tasks, like editing texts, preparing budgets, storing and searching information, drawing graphics and tables, performing calculations and an infinity of other things. Nowadays, software applications are no longer limited by hardware-related aspects, but instead by the human imagination and by cost restrictions and user habits (Cusumano 2004, pp. 280–281).

> "The aim of marketing is to know and understand the customer so well the product or service fits him and sells itself."
> *Peter F. Drucker (1909–2005), management consultant*

There are some aspects that distinguish the tailor-made software systems from the software products, namely those related to the origin of the necessity that led to their development. That necessity may come from an individual person or from the market (Wieringa 1996, p. 34). In tailor-made software systems, the origin is clear, since development occurs after an explicit manifestation of the necessity, made by the potential client, to support the realisation of individual tasks or the execution of organisational processes. The development effort related to software products can be initiated, regardless of an explicit manifestation of the necessity by the potential clients. It is common that the necessity is identified by marketing experts, based on market studies and analysis of consumption trends. For tailor-made software systems, only one unique instance is usually made available, whilst for software products it is necessary to produce various installations for exploration by different clients. The software industry is a business area with a high return on investment, where making a unique copy of a software product or several thousands costs roughly the same (Cusumano 2004, p. 15). This reality allows the investments in software to have the potential to result in substantial profits, as also happens in the film, music, and pharmaceutic industries. However, the development of a software product has normally a relatively high fixed cost that is not recoverable, if the product is not a commercial success.

In some cases, it is difficult to classify a software system. For example, ERP (*enterprise resource planning*) software products, like the SAP ERP or the Microsoft Dynamics NAV, are generic, but can be configured to respond to specific requirements of a given client. Despite the referred terminology, the differences between a software system, a product and an application are often not significative, which means that the three terms are practically used as synonyms. A tailor-made software system can be transformed into a software product, through the generalisation of the functionalities it offers. The contrary is also true, that is, a software product can be adapted to satisfy the particular requirements of a given client. For this reason, in this book, the three terms are, in many situations, used for representing software artefacts, with high complexity and whose development requires approaches, methods, and tools from the software engineering sphere.

2.3.3 Domains

Engineering systems are developed due to the existence of some necessity of the stakeholders that must be satisfied. The area in which the system is explored is designated as domain. It is necessary to precisely characterise what is its meaning, since this word has different meanings, as a function of the context where it is used. Generically, a domain can be considered as a business area, collection of problems, or knowledge area with its own terminology. Within this book, a **domain** is an area of human knowledge or activity that is characterised by possessing a set of concepts and terms that the respective players know.

Examples of domains are telecommunications, transports, health, agriculture, industry, retail, banking, insurance, education, entertainment, cinema, and theatre. Domains can involve the physical world (for instance, a library involves the manipulation of books) or can be intangible (for example, schedule management). Generally, the domains have no relation with computers, although there are exceptions (for example, hardware commerce or source code management). The domains where the software technology can be present are only limited by the human imagination.

It is not obligatory that, before initiating the development of a system in a given domain, the requirements engineer has any knowledge about that domain. Obviously, it is desirable that she is at least comfortable with some of the basic concepts of the domain, so that he can speak in an comprehensible way with the system stakeholders. Over the project, the requirements engineer must increase the knowledge that he possesses about the domain, even as a mechanism to make sure that the development team takes into account the client's perspective. Some software producers are specialised in a given domain, in order to gain a competitive advantage with respect to their competitors. Although this approach reduces the potential market, the solid knowledge on the respective domain permits more specialised systems to be developed, thus offering a better answer to the users needs.

In relation to a given system, it is common to refer to two distinct domains: the problem domain and the solution domain. The *problem domain* is the context where one feels the necessities that need to be satisfied by the system to be developed. For instance, in the case of a restaurant, the problem domain includes the elements that characterise it: clients, cooks, tables, chairs, towels, cutlery, crockery, meals, etc. If the problem domain is an airline company, then one can see, for example, airplanes, pilots, stewards, passengers, suitcases, and tickets. The persons have technical or business problems that can be solved with the engineers contribution. The aim of the requirements engineers consists in understanding what are the problems of those persons, in their language and culture, so that one can construct systems that satisfy the necessities of those persons (Leffingwell and Widrig 2000, p. 19). The *solution domain* refers to the activities that are executed and the artefacts that are handled and constructed to solve the problem.

Davis (1990, pp. 29–32) classifies the software systems domain along five orthogonal axes:

1. difficulty of the problem class;
2. temporal relationship between data and processing;
3. number of tasks to be executed simultaneously;
4. relative difficulty of the data, interaction, and control aspects of the problem;
5. determinism level.

> "If there is no solution to the problem then don't waste time worrying about it. If there is a solution to the problem then don't waste time worrying about it."
>
> *Dalai Lama XIV (1935–)*

The problem class is about how the real problem (i.e., the problem felt by the stakeholders in the context of the problem domain) can be framed into the conceptual problem that covers it and that may have been previously studied and analysed by experts. For example, the real problem of a logistics company that aims to define routes that minimise the delivery times and fuel expenses can be framed into a more conceptual problem, like the travelling salesman or Chinese postman problems. The difficulty of the problem class can be divided into two groups. The *difficult problems* are those that were never solved or that do not have any satisfactory solution. The *not-difficult problems* are all the others, that is, those that have been previously resolved in a reasonable way.

With respect to the temporal relations that exist between the availability of the input data and its processing, there are also two classes. In *static applications*, all the inputs must be available before the application processes them. In *dynamic applications*, the input data arrive continuously during the processing, therefore having an effect on the results. Compilers are typically static applications, while interactive systems are examples of dynamic applications.

A third alternative of classification is related to the number of tasks that the system can simultaneously handle. *Sequential applications* manipulate a single task at a time, while *parallel applications*, from the users' perspective, must be capable of processing several tasks in simultaneous. The most difficult aspect related to the externally observable behaviour of the system to address constitutes another classification axis. It includes three cumulative dimensions that can be considered: data, interaction, and control. In a *data-centred application*, the type, organisation, and persistency of the data that support that application are the most critical aspects to consider. In *interactive applications*, the most difficult aspect to handle is how the environment and the system interact by exchanging and presenting information. In applications with a strong *algorithmic component* or *decision taking*, the primary aspect is the relation that is established between the system inputs and outputs, forcing or permitting the control levels caused by the exchange of information with the environment. These three alternatives are not mutually exclusive, since it is possible to observe significant manifestations of these three types of behaviour in the same system.

A last axis of classification refers to the predictability of the system outputs as a response to the inputs. In *deterministic systems*, one expects the same results to be produced for the same set of inputs. For example, a scientific calculator must always provide the same result for the same inputs. *Non-deterministic systems* provide answers that are not absolutely clear. It is possible that different outputs can be accepted as valid. For example, a software application to play chess can, in each move, opt for any of the various valid alternatives. There will be certainly some moves that are better than others, but that evaluation has not a unique answer, in the generality of the cases.

2.4 Models for the Development Process

The software process can be executed in different ways and according to different approaches. A *process model* represents a development process and indicates the form in which it must be organised. The process models aim to help the engineers in establishing the relation among the activities and the techniques that are part of the development process. Whenever the development process is modelled, one can reap the benefits that result from the systematisation and identification of the best practices, to allow systems development in an efficient, reliable, and predictable way. With the development process systematisation, through the definition of the respective model, one tries to reach the following objectives:

- to clearly identify the activities that must be followed to develop a system;
- to introduce consistency in the development process, ensuring that the systems are developed according to the same methodological principles;
- to provide control points to evaluate the obtained results and to verify the observance of the deadlines and the resources needs;
- to stimulate a bigger reuse of components, during the design and implementation phases, to increase the productivity of the development teams.

The difficulties faced by the software development teams and, more specifically, their managers lies in defining processes that promote the utilisation of management mechanisms that keep the projects under control. The challenge here is not blocking the necessary creativity and flexibility that is required so that the system at hand can adapt to changes both in technology and the users needs.

The next subsections present and characterise the most fundamental process models. To completely describe a process, several facets should be considered: the activities (set of tasks that must be executed to develop the system), the artefacts (results of the activities), and the roles (responsibilities of the persons engaged in the process). In the majority of the cases, this book presents, in a graphical form, only the activities, leaving for the textual part the discussion of the other facets whenever necessary.

Fig. 2.2 The waterfall process model

2.4.1 Waterfall

The oldest software development process model is designated as the *waterfall model*. As Fig. 2.2 depicts, it is composed of a sequence of phases, namely analysis, design, implementation, and testing. The use of the 'waterfall' word aims to evince the irreversibility whenever one progresses from one phase to the next one, as well as the risk associated with the process execution. The most relevant characteristic of this process model is the strong tendency for the development to follow a top-down approach (from the most abstract to the most concrete) and, in a high-level perspective, the strictly-sequential progression between consecutive phases (Yourdon 1988, pp. 45–47).

During the *analysis* phase, the functioning of the system is specified, through the identification of the various requirements that must be considered. The document that contains the specification serves as a basis for the next phases, so, ideally, one should use implementation-independent notations and allow all stakeholders to clearly understand (i.e., without any ambiguities) what are the intended functionality.

Once the document that specifies the system under development is accepted, the *design* phase, which consists in transforming a specification into an architecture, begins. According to Bosch (2000, p. 230), the most complex activity during software development is precisely the transformation the requirements into an architecture. Generically, this phase is divided into two steps. The first one, designated as architectural design, describes how the system is constituted and is, in many cases, one of the most creative tasks in all the development process (Stevens et al. 1998, p. 88). In this first step, an architecture must be established, by identifying the system components and possible restrictions in their behaviour. This architecture determines the internal system structure, defined based on the entities that compose it and in the relations among them. Whenever the architecture is defined, the second step, designated as detailed design, establishes in detail the components, in order to include enough information to allow its implementation. Sometimes, one considers the existence of a step, called the mechanistic design, that relates the decisions taken in the architectural design and in the detailed design, through the detailed study of the mechanisms that provide the architecture with the expected behaviours for the system.

In the design phase, the principal objective is to structure (i.e., to define the architecture of) the system at hand. For example, an object-oriented design includes the object-oriented decomposition process, using an appropriate notation to describe all the (logical or physicals and static or dynamic) aspects related to the system (Booch et al. 2007, p. 42). In a system conceived according to the object-oriented paradigm, the respective structure is dictated by the objects that compose it and the relations that are established among them.

The principal difference between the analysis and the design phases is that while the former produces an abstract model that mimics the fundamental aspects of the existing needs in the problem area, the latter creates a model that specifies the components that structure a particular system solution. In other words, the analysis phase defines the system functionality (what to do), while the design phase stipulates the architecture (how to do) that the system must present so that the expected behaviour is obtained.

Despite the many methodological and technological advances that have occurred in the last years, the software is often developed in an handicraft way. The software industry is still far from reaching a point where there is, in fact, an entire catalogue of software components able to be easily and directly integrated in the systems. In software engineering, the development of all the parts of a system is not only generally accepted but also often encouraged, with the argument that this is the only way to build the system with the intended quality. This argument is called the not invented here (NIH) syndrome Lidwell et al. (2010, pp. 168–169). Such scenario would be, nowadays, unthinkable in other industries. Fortunately, there are many mechanisms, such the component libraries, frameworks, application programming interfaces (APIs), design patterns, which provide very significant advances in this area. Web applications constitute an important niche, where the use of reusable components, like web services, is quite common.

The *implementation* (codification or programming) phase transforms the models defined in the design phase in executable code. This transformation involves the definition of the internal mechanisms so that each component can satisfy its specification and the implementation of those mechanisms with the chosen programming language (Zave 1984). The implementation phase is considered by many authors, for instance, Hatley and Pirbhai (1987, p. 10), Rumbaugh et al. (1991, p. 278) and Whytock (1993), as a a purely mechanical, simple, and direct task, after the most intellectual and creative work has been performed in the analysis and design phases. The implementation phase is thus a serious candidate to be automated, if one can rely on tools that permit us to indicate how the final code can be generated from the specifications obtained in the previous phases. However, reality has shown that it is not always so easy to pursue with this phase. Object-oriented programming allows systems to be implemented, organised as collections of objects. Each object is an instance of a class and each class is a member of a structure where there are hierarchical relationships.

While in the development of non-complex systems, the effort devoted to the analysis and design phases can be residual (when compared with the effort in the implementation phase), for complex systems the effort devoted to issues related to analysis and design is of vital importance. Actually, the popularisation of tools, which automatically produce code from specifications, allows one to say that the point where "the specification is the implementation" is about to be reached. The essence of developing systems becomes thus focused on decisions related to the analysis and design phases.

Even when the engineers carefully follow development processes and approaches, the artefacts that result from development activities may still contain defects. To avoid these possible defects, the systems must be tested. *Testing* has two main objectives: (1) to show that the system under development does what it is expected to do, and (2) to allow defects in the system to be discovered.

> "If debugging is the process of removing software bugs, then programming must be the process of putting them in."
>
> *Edsger W. Dijkstra (1930–2002), computer scientist*

The term 'testing' encompasses a very large set of activities that go from testing a small piece of code by the programmer (unit testing) up to the validation, by the client, of a software system (acceptance testing). To distinguish some of these activities, the community adopts the terms 'verification' and 'validation'. By *verification* one means the process through which it is ensured that the system was built in accordance with the requirements and the specifications. The verification can be achieved through dynamic approaches, in which the run-time behaviour of the system is checked, or static ones, where one analyses and inspects any system-related artefact or document (Ghezzi et al. 1991, p. 260). The aim of *validation* is to make sure that the system satisfies the necessities and expectations of the users and the clients.

The testing phase was traditionally executed at the end of the development process. The program code was totally written, before executing any function testing. However, this vision needed to be changed, as soon as it was clear that testing is more that just debugging code. Software testing, if well performed, can result in huge economical benefits. Nowadays, testing complex software takes around 40 % of the development costs (Ebert and Jones 2009; Sommerville 2010, p. 6), which demonstrates the growing importance that it has in the software engineering context. Actually, the success of software testing depends on the planning of its execution and its effective realisation in the initial development phases. If quality of a software system is strongly dependent on the quality of the adopted development process, similarly, the quality and effectiveness of testing are largely determined by the quality of the testing process (Kit 1995, p. 3). Software testing has its own lifecycle, which is realised at distinct levels: it starts at the same time as the requirements elicitation and, from that point on, follows in parallel with the development process. In other words, for each phase or activity of the development process, there is an associated testing activity, as Fig. 2.3 depicts (Robinson 1992, p. 3). This figure, which represents the V process model, can be viewed as a refinement of the waterfall model shown in Fig. 2.2.

The first level of testing (unit testing or unitary testing) takes place as the diverse components (units) are implemented. The objective is to verify if each component in isolation works as expected, using the decisions taken in the detailed design as reference. When the components pass the unit tests, the following step is the *component integration testing*, whose purpose consists in guaranteeing that the interfaces between the components have the behaviour estimated during the architectural

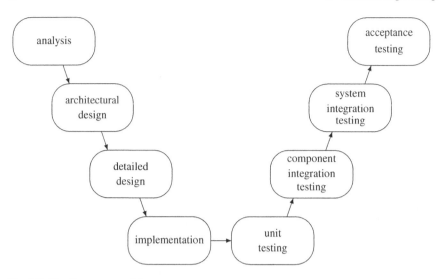

Fig. 2.3 The V process model

design. Next, the *system integration testing* permits one to verify if the software system, as a whole, satisfies the requirements indicated in the analysis phase. The two types of integration tests indicated can be very demanding activities in terms of resources and time, especially for critical systems. Finally, the *acceptance testing* is executed jointly with the end users that validate the operation of the system with respect to their expectations, in the light of what has been contracted. Some testing activities can be automated, partially or even totally, and that possibility depends on the notations used in the analysis and design phases.

All these phases (analysis, design, implementation, and testing) are related among them and none should be neglected during the development of a given system. However, the division among these phases is not always so explicit as it was indicated up to now. For example, in the object-oriented methodologies, there is normally an overlap in the tasks covered in the analysis and design phases, since the separation between those two phases is more theoretical than real, being very difficult to define the respective frontier (Booch et al. 2007, p. 131). This fact can be interpreted as a disadvantage, but a positive perspective is also possible, meaning that the transition between those phases is made in a natural and soft way, when an object-oriented decomposition is followed.

The waterfall model, due to its conceptual simplicity is still one of the most referred software development processes, despite the various problems that it presents and the many alternatives that were proposed. The waterfall model is considered too inflexible and produces satisfactory results only when the requirements are clear and the chances of being changed too low. To develop, for instance, a compiler, based on a grammar completely defined and which is not likely to be changed, this model seems to be perfectly adequate (Ghezzi et al. 1991, p. 374).

However, the waterfall model, is based on documentation as a criterion to decide when a phase is finished and the next one can start, which requires complete documents to be written. This fact is extremely negative, in projects where the ideas and the necessities of the stakeholders are not yet totally clear, since it forces all the requirements to be decided very early. This fact probably introduces errors that will be spread to the design and implementation phases. Unfortunately, those errors are only detected when the implementation is finalised, usually quite after the specification of the requirements.

The processes that follow the waterfall model tend to be accomplished according to a plan in which generically the requirements are specified completely, so that one can subsequently design, construct, and test the system. The waterfall model, despite allowing the process to be executed in various iterations, is essentially characterised by following a process with a significant level of bureaucracy and ceremony.

In practice, the order in which the phases are executed is difficult to capture precisely, because it is not necessary that a phase finishes totally for the next one to be initiated. This fact is contrary to what the waterfall model theoretically proposes. Usually, the results that are obtained in a given phase are used for correcting the results of the previous phases. Thus, the processes should be iterative in practice, since problems identified in more advanced phases of the project force the previous phases to be revisited.

2.4.2 Incremental and Iterative

In development contexts, where the market is extremely accelerated and vibrant, with new opportunities arising at very fast rhythms, the waterfall process is very inadequate. The *incremental and iterative model* is based on the characteristics of the waterfall model, introducing however iterations to permit an incremental development. As Fig. 2.4 illustrates, this process model applies linear sequences of development in a phased way. Each linear sequence produces as result a functional increment of the system. For example, the construction of a text editor could proceed through the following iterations:

- in the 1st iteration, one develops the functions to manipulate files (open, save, close, print) and the basic functionalities for editing text (insert, delete, select);
- in the 2nd iteration, one includes more advanced edition capabilities (search and replace text, fonts, bold, italics);
- the 3rd iteration allows the inclusion of figures, tables, and graphics in the documents, by providing new commands that manipulate those elements;
- in the 4th iteration, new functionalities that permit the use of thesaurus and automatic spellers are added.

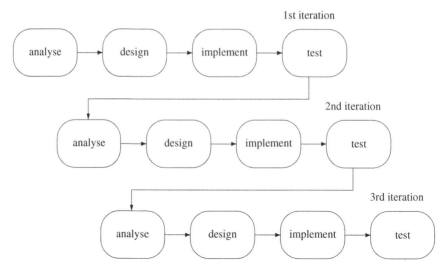

Fig. 2.4 The incremental and iterative process model

"You should use iterative development only on projects that you want to succeed."

Martin Fowler (1963–), software engineer

The incremental and iterative model is based on the idea that it is easier to create a simple artefact than a complex one and also that it is simpler to modify an existing artefact than to create a new one from scratch. Hence, instead of trying to completely construct the system in a unique cycle, the attention at the beginning is deliberately focused on incorporating the most critical and important aspects (or those that are more clearly mastered). Afterwards, new aspects can be progressively incorporated, until the system is complete.

The incremental and iterative process is, in its essence, repetitive, i.e., the functionalities are gradually added and improved until the system is fully handled. Each increment represents a part of the system functionality.

Agile methods for developing software can be framed within this type of process model. They appeared, in the 1990 decade, as an alternative to the traditional methods, dismantling many of the assumptions associated with the latter. The accelerated dynamics in various domains and businesses make impossible, from a practical point of view, obtaining a set of stable requirements, since they are subject to constant and unexpected changes. Agile methods seek to position themselves as an adequate alternative for these highly unstable scenarios. For that purpose, they follow four fundamental principles, established by Beck et al. (2001), which, despite recognising value and utility to the items on the right, give prevalence to those on the left:

individuals and interactions	vs.	processes and tools
working software	vs.	comprehensive documentation
customer collaboration	vs.	contract negotiation
responding to change	vs.	following a plan

Agile methods aim essentially to minimise the risk associated with software development, defining for such very short development cycles, called iterations (or *sprints*). They are incremental and iterative development methods, in which the cycles last between one and four weeks. An agile software project aims to produce, at the end of each iteration, a new version/release of the system, that is, a version that is executable, that works correctly and that provides value to the client. Hence, each iteration includes a cycle with all the tasks necessary to concretise the inclusion of the new functionalities: requirements engineering, design, implementation, testing, and documentation.

> "What is the difference between method and device? A method is a device which you used twice."
>
> *George Pólya (1887–1985), mathematician*

At the end of each iteration, the team reevaluates the requirements and can introduce alterations. This practice permits that set of requirements to be changed, through the inclusion of new requirements and the elimination or change of requirements previously identified. Additionally, the priority of each requirement can be changed. This reevaluation of the requirements implies that they can be effectively changed, during the project, thus increasing the utility and value of the system for the stakeholders.

Agile methods lie in the real-time and, if possible, face-to-face collaboration among the development team, the clients and the users. This collaboration allows the team to discuss the project scope, analyse and prioritise the requirements, and decide the options to be taken. Thus, it is not so critical the existence of written documents to support the development tasks. Actually, most agile software projects, in comparison with other approaches, produces less voluminous documentation.

2.4.3 Transformational

Although they are not yet a generic solution for all types of systems, mathematical methods (typically designated formal methods) offer the very attractive perspective of generating software with no defects. The use of formal methods presupposes a *transformational process* that assumes the existence of tools that automatically convert a formal specification into a software program that satisfies that specification. This process implies that the changes are reflected in the specification, thus eliminating the problem of getting *spaghetti code*, i.e., code that gradually becomes poorly structured, as a consequence of being successively modified.

Fig. 2.5 The
transformational process
model

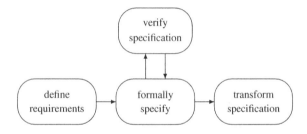

Figure 2.5 illustrates the transformational process model that is composed of four main tasks. In the first task, requirements are elicited, with a set of techniques that are considered adequate. Based on the requirements, a formal specification is created and is progressively developed, until an executable version is obtained. This characteristic means, in this context, that the specification can be processed, from which results a program that can be executed. An executable specification is more that the mere static model that defines the system behaviour and should allow the verification of the behavioural properties. Some of the benefits observed in the utilisation of executable specifications, in several large-scale projects, are described by Harel (1992). If some non-conformity is detected during the verification, the specification should be modified, in order to eliminate that problem, and verified again. This cycle repeats the necessary number of times until the specification is in accordance with the requirements.

> "It is easier to change the specification to fit the program than vice versa."
> *Alan J. Perlis (1922–1990), computer scientist*

The specification is used to obtain, through automatic transformation mechanisms, parts of the final program with the efficiency levels specified in the requirements. The process of transforming the specification is controlled by the software engineer, which thus can vary the characteristics of the program, taking into account, for instance, the non-functional requirements. This process model becomes efficient and useful, if there is an environment that provides tools to automatically support the various activities, especially those to transform the specifications into the program.

The specification changes are considered as a part of the process. Contrarily, in a waterfall process, changes are seen as reparations, since their occurrence is not considered beforehand. From here it results that changes are made under big pressure, normally at the end of development, implying often that the changes are made modifying directly the code, without reflecting them in the specification. Hence, the specification and the implementation diverge one from the other, making any future changes more difficult to accomplish. This situation does not occur in a transformational process, since the process executed to develop the software system and the respective decisions (intermediate steps, based on mathematical proofs) for each transformation are registered, which makes it possible to restart, from an intermediate point, the transformation of the specification into an implementation.

Despite the growing maturity that the formal methods have acquired, their utilisation is not yet disseminate in a pervasive way, being its application restricted to well-identified areas. Among the disadvantages of their utilisation, one can find the long development time, the difficulty in using the specifications as a medium of communication with the clients, and the need to resort to specialists to manipulate the mathematical specifications. This last argument should not be taken literally, since some formal methods employ concepts familiar to any engineer. More recently, the model-based development approach resort to the transformational principles of the formal methods, but it adopts models closer to the implementation languages (than the strictly-mathematical approaches), as a way to make viable the transformational approach in large-scale real projects.

2.4.4 Spiral

The *spiral process model* (Boehm 1988) is based on a risk-driven approach and not on documents or the code. In this context, a **risk** is a potentially adverse circumstance that can have negative or perverse effects in the development process and in the final quality of the system. A risk is a measure of the uncertainty of achieving an objective or meeting requirements. The spiral model centres its action in the identification and elimination of problems with high risks.

The various tasks are organised in cycles, as Fig. 2.6 documents. Each cycle of the spiral is constituted of four main tasks, being each one represented by a quadrant of the diagram. The radius of the spiral represents the progress in the process and the angular dimension indicates the accumulated cost in the process.

In the first task of the cycle, one identifies the objectives (performance, functionality, easiness of modification, etc.) for the system under development, with respect to the quality levels to be achieved. It is also relevant to identify the alternative means of implementation (develop A or B, buy, reuse, etc.) and the restrictions that must

Fig. 2.6 The spiral model

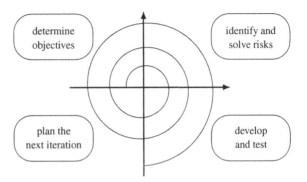

be assumed to materialise one of those alternatives. In the second task of the cycle, one evaluates the alternatives previously identified with respect to the objectives and restrictions, which frequently implies the identification of uncertain situations that represent potential sources of risk. To proceed with this identification, one can resort to different techniques, like prototyping, benchmarking, simulation or questionnaires. During the third task, one should develop and verify the system for the next cycle, based again on a risk-oriented strategy. In the fourth and last task of the cycle, the obtained results are reviewed and the next spiral cycle, if that is the case, is planned.

> "The basic problem of software development is risk."
>
> *Kent Beck (1961–), software engineer*

Whenever the requirements of a system are reasonably well known, a waterfall process can be followed, which means that only one spiral cycle is fulfilled. For systems whose requirements are less clear, several cycles may be necessary to achieve the intended results, from which results an incremental and iterative process.

As special cases, the spiral model includes the process models presented in this section. It allows the choice of the best combination and composition of those process models for each situation in which it is applied. Therefore, the spiral model can be seen as a meta-model, i.e., a reference model to instantiate different process models.

2.5 Summary

Software engineering is an engineering discipline focused on all the aspects related to the development and production of software. The software engineer, as any engineer, is responsible for the development process and production of artefacts that are to be used by third-parties. Software engineering addresses the development and production of software systems made by teams constituted by several professionals. This implies that software engineering is not just centred in the technical aspects associated with software development, but includes also process management activities. Software engineering brought to the computer science field the junction of the development processes and methods with economic and management issues. These questions are indispensable for the professionals that actuate in the industry with roles and responsibilities that go behind the mere computer programming.

The software engineering body of knowledge is structured into 15 KAs according to the SWEBOK guide. Software engineering is a scientific area with an extremely important economic and social relevance and that has allowed people to observe many of their daily tasks solved in a faster, more comfortable, and more economic way. This success benefits from the capacity of software engineering to adapt to the enormous demands that nowadays their professionals are subject to. They have to handle all the software production process issues, but simultaneously possess the technological competencies and sensibility for the necessities and expectations of the users. The software engineering essence lies upon the capacity to put in practice the most appropriate sets of cooperative, coordinated and concurrent technological and methodological approaches for each software development reality.

Software is the principal artefact that results form the software engineering process. The chapter presents the most relevant software characteristics and defines and presents the various types of software systems. The software systems are developed to satisfy the stakeholders needs. The domain of a software system can be a business area, a collection of problems, or a knowledge area.

The chapter ends with a presentation and characterisation of the most popular process models used to guide the development teams in the organisation of their tasks. There is no unique development process model that is adequate for all the projects, which means that it is very important to choose the one that best adjusts to each particular context.

Further Reading

The literature in software engineering is quite extensive. There are some books that try to make a complete coverage of all the discipline. To start with, there are some works considered as true classics, like the encyclopedia-like books written, over the past years, by Sommerville (2010) and Pressman (2009). These books have go through several editions (always with updates in relation to the previous version), in the case of Pressman every 5 years approximately, which demonstrates the highly accelerated dynamics of this discipline. Another interesting book, written by Pfleeger and Atlee (2009), focuses on modelling and agile methods. Another landmark book in this area is the work authored by Ghezzi et al. (1991), which, despite its generalist nature, is quite concise and easy reading.

The history of software is also very interesting. Two recommended sources are Ceruzzi (1998, Chap. 3) and Campbell-Kelly (2003). Broy and Denert (2002) edited a book with a compilation of some of the scientific articles that, since the 1950s, changed somehow software engineering. This book constitutes a compilation of undeniable rigour and historical value, even though one can contend that "the" software engineering pioneers are not necessarily just the authors included in the book. An extremely interesting essay about the differences between software engineering and other more traditional engineering fields can be found in Beizer (2000).

Software engineering education is addressed by various authors, for example, Parnas (1990, 1999), Shaw (1990, 2009), Osterweil (2007). It is also relevant to read the *curricula* recommendations for software engineering (ACM/IEEE-CS 2004, Pyster 2009).

In the field of the software process, it is recommended that one reads the works about the software development process improvement, like, for instance, (Humphrey

2005). The book written by Krutchen (2003) presents the unified process. With respect to agile methods, it is suggested that one reads books related to eXtreme Programming (XP) (Beck 2000) and Scrum (Schwaber and Beedle 2001).

Exercises

Exercise 2.1 Indicate the 15 KAs that constitute the software engineering body of knowledge, according to the SWEBOK guide.

Exercise 2.2 Identify which elements are included in a software product.

Exercise 2.3 A *mobile application* is a software application, developed to run on a smartphone or other handheld device. Identify the most important characteristics of a mobile application, according to Salmre (2005, Chap. 2).

Exercise 2.4 (Naveda and Seidman 2006, pp. 23–24) While developing a software application, two similar defects were detected: one during the requirements phase and another one during the implementation phase. Which of the following sentences is more likely to be true?

1. The most expensive defect to repair is the one detected in the requirements phase.
2. The most expensive defect to repair is the one detected in the implementation phase.
3. The repair cost of the two defects tends to be similar.
4. There is no relation between the phase in which a defect is detected and the repair cost.

Exercise 2.5 Point out some advantages that result from using an incremental and iterative process in comparison with a sequential process that follows the waterfall model.

Chapter 3
Requirements

Abstract A requirement identifies an attribute, a capacity, a characteristic or a quality that a system should exhibit in order to have value for the users and customers. This chapter defines what is a requirement and presents the different types of requirements (functional requirement, non-functional requirement, user requirement, system requirement). The chapter ends with a discussion of several concepts related to requirements, such as functionality, use cases, service and feature.

3.1 Definition of Requirement

Colloquially, a requirement can be defined as anything that someone desires. In the context of systems development, requirements are seen as properties that the systems (yet in project) shall possess when built. The requirements express the users necessities and restrictions that are placed on the system and that must be considered during the development. From the point of view of the systems engineer, a requirement can also be defined as something that needs to be conceived. There are many more definitions, but here it is relevant to present the definition for requirement proposed by the IEEE 610.12-1990 standard (*glossary of software engineering terminology*):

1. a condition or a capacity that someone needs to solve a problem or to achieve an objective;
2. a condition or a capacity that must be verified or possessed by a system or by a system component to satisfy a contract, standard, specification, or other formally imposed documents;
3. a documented representation of a condition or capacity, the in (1) or (2).

There are many ways and dimensions for classifying the requirements. However, those classification mechanisms should not be seen as absolute truths, since they are not totally separated or orthogonal among them. Furthermore, in some cases, a given classification mechanism may not be easily applicable to some requirements. It is important to say that this fact is also not critical, since, more important than classifying with great detail and precision a given requirement is to identify that it exists and to understand its meaning.

© Springer International Publishing Switzerland 2016   45
J.M. Fernandes and R.J. Machado, *Requirements in Engineering Projects*,
Lecture Notes in Management and Industrial Engineering,
DOI 10.1007/978-3-319-18597-2_3

The IEEE 610.12-1990 standard definition introduces a first classification for the requirements, dividing them into two alternatives. One related to the users needs and another one with the system capacity. These two alternatives are strongly associated with two different requirement types that are designated as user requirements and system requirements (discussed in detail in Sect. 3.4). The third alternative of the definition has, in our opinion, a different nature, in accordance with what is discussed by Kaindl and Svetinovic (2010), since it mixes the concept of requirement with its representation. Succinctly and providing support to the two first alternatives in the definition, one can consider a **requirement** as a capacity that a system must possess, to satisfy the users necessities.

> "I don't think necessity is the mother of invention—invention, in my opinion, arises directly from idleness, possibly also from laziness. To save oneself trouble."
>
> *Agatha Christie (1890–1976), novelist*

Requirements representations should display a series of properties. Each requirement must be clear, that is, easy to interpret, implying, in the cases where it is written in a natural language, the use of simple sentences. One expects also that each requirement is not ambiguous, i.e., it only allows one interpretation. These two characteristics can be obtained by concisely writing the requirement, without using technical jargon or acronyms. The requirement should express objective facts and not personal opinions. These two characteristics are hard to ensure, but some of the guidelines presented in Sect. 7.1 address these difficulties.

It is also possible to divide the requirements between functional requirements (Sect. 3.2) and non-functional requirements (Sect. 3.3). Classifying a requirement as either functional or non-functional depends on the point of view of the observer. A requirement that for a stakeholder can be perceived as being functional, can for a different stakeholder be considered as non-functional. For example, the form of a skyscraper is a non-functional requirement for the civil engineer, but can be a functional requirement for the urban architect, if the construction of that skyscraper aims to serve as a touristic attraction of the city (Pohl 1996, p. 4).

Beyond these two principal forms of classifying the requirements, there are other alternatives. For example, Aurum and Wohlin (2005b) distinguish primary and derived requirements. A **primary requirement** originates directly from some stakeholder, i.e., it was requested by that person or entity, while a **derived requirement** is obtained by refining a primary requirement.

The word 'requirement' can also mean an indispensable condition or demand. For instance, in the context of an application for a job position, a requirement is something obligatory, that is, a prerequisite. However, the requirements within the scope of an engineering project are negotiable. It is normal that a requirement, initially perceived as important, to have its importance decreased after the client becomes aware about the respective cost. A **candidate requirement** is a requirement that was identified by some elicitation technique. These candidate requirements must take into account the intentions with respect to the system from the clients. Their incorporation in the system depends on the agreements that are established in the negotiation process. The use of the term 'candidate' aims to emphasise the possibility of the requirement not being considered.

> "The list of requirements for the design phase is often 50 times longer than the list of original requirements."
>
> *Robert L. Glass (1932–), software engineer*

3.2 Functional Requirements

A **functional requirement** describes a functionality to be made available to the users of the system, characterising partially its behaviour as an answer to the stimulus that it is subject to. This type of requirements should not mention any technological issue, that is, ideally functional requirements must be independent of design and implementation aspects. Consequently, the solution space is as broad as possible, henceforth increasing the technological alternatives that can be explored during the project.

In a collective perspective, the set of functional requirements must be complete and coherent. The set is *coherent* if there are no contradictions among its elements and *complete* if it considers all the necessities that the client wishes to see satisfied. These two characteristics are hard to ensure, especially for highly-complex systems. Davis (1990, p. 188) considers that defining complete requirements is the most difficult attribute to achieve or evaluate, since it is hard to detect if something is absent by examining what is present.

The requirements that are obvious, due to several types of impositions or by the practices in the domain, are frequently forgotten and as such are not documented, neither negotiated during the development process. In the circumstances in which those (obvious) requirements should be undoubtedly implemented in the final system, the analyst must guarantee that they are documented and correctly handled by the development team. Since the client may not be aware about these requirements, they are commonly referred to as implicit requirements. An **implicit requirement** is a requirement included by the development team, based on the domain knowledge that it possesses, in spite of not having been explicitly requested by the stakeholders.

In some cases, it can be relevant to indicate what is expected the system <u>not</u> to do, to avoid implementing the implicit requirements against the will of the stakeholders. A classic example of an implicit requirement is the inclusion of an entrance door in a house. In the architecture team, no one is expecting that the client asks for an entrance door, neither that the house does not include that door, even when the client does not request it. Hence, if in a given context someone wants an house without entrance door, that "non-requirement" must be explicitly requested. In software, it is very common that web applications have user authentication (for example, through the indication of a username and a password), even if the stakeholders have not requested that functionality. The same happens, for example, with the graphical interface that one expects to be as friendly as possible. On the contrary, an **explicit requirement** refers to a requirement that was requested by the clients and that is represented in the documentation. In reality, these are transitory properties, since when an implicit requirement is referred to it becomes immediately explicit functional requirement.

3.3 Non-functional Requirements

A **non-functional requirement** corresponds to a set of restrictions imposed on the system to be developed, establishing, for instance, how attractive, useful, fast, or reliable it is. Classic examples of this type of requirements are time constraints, restrictions in the development process or adoption of standards. In fact, it is common that some projects include a requirement like, for instance, "the product must be developed in C++". This example represents a technological restriction that must be followed throughout the product development. It is important that the relevance of a requirement of this type is discussed and agreed upon between the clients and the development team to avoid taking design and implementation decisions prematurely and hastily.

It should be clear that the expression *non-functional*, when applied to a requirement, does not aim to mean a requirement that is not capable of performing a given function (Chung and Prado Leite 2009). To avoid that possible interpretation, some authors suggest, in alternative, the use of the terms 'quality requirement' or 'quality attribute'.

A non-functional requirement does not change the essence of the system functionalities. For example, in sports the colour of the ball does not affect its functionality, but it is probably a bad choice to use a white football in a field full of snow. Generically, the functional requirements remain the same, regardless of the non-functional requirements associated with the system. So, non-functional requirements and functionality are orthogonal, in the sense that many software designs could achieve the same functionality with distinct qualities. Therefore, software architects focus more on quality attributes rather than on functionality (Fairbanks 2010, p. 142). It is normal that some projects have their origin precisely in the non-functional aspects; if the current system is slow, unreliable, or difficult to use, then these characteristics may lead to the creation of a better system that does exactly the same, but in which the non-functional requirements gain prime importance.

One of the difficulties that is observed when handling non-functional requirements results from the fact that those requirements are applicable to the system as a whole and not just to some of its parts. This means that generally non-functional requirements cannot be modularised. For example, if one wishes the system to be inexpensive, that characteristic has a global impact on the system as a whole, being impossible to ensure its fulfilment just by a subset of its components. In this regard, the property of "being inexpensive" must be expressed by a numerical form indicating the value in a given currency (for instance, Euros or Dollars). In this sense, the non-functional requirements are frequently emergent properties of the system at hand. An **emergent property** of a system cannot be determined solely by the properties associated with its components, but is additionally determined by the system structure, i.e., how the components are interconnected to form the system (Thomé, 1993a). An emergent property of a system is a property that can be associated with the system as a whole, but not individually to each of its components. Reliability is a good example of an emergent property of a system. Actually, it is not sufficient that all components are reliable, for the respective system to be also reliable. The size of a software application, measured by the number of bytes, is a example of a property that is not emergent.

If the system is designed only based on the functional requirements (i.e., without considering the non-functional requirements), it may exist as a monolithic entity, with no internal structure. Non-functional requirements are crucial to decide the system architecture, and so handling those requirements is mandatory in the project of engineering any system (Illustration 3.1). The fulfilment of a non-functional requirement cannot be achieved in an isolated way, that is, one cannot maximise a given

Illustration 3.1 Automobiles are different according to the functional and non-functional requirements that they satisfy

non-functional requirement without sacrificing some other non-functional requirements. The selected level for the satisfaction of a given non-functional requirement affects, either positively or negatively, the satisfaction of other non-functional requirements. For example, a system optimised for performance can see a reduction in its characteristics associated with maintainability. Adaptability, for instance, contributes positively to portability.

There are some proposals to classify non-functional requirements. For example, the classification scheme suggested by Sommerville (2010, p. 88) divides in a first level the non-functional requirements into three categories:

- **product requirements**: characterise aspects of the behaviour of the system itself, including, for example, (i) reliability, (ii) performance, (iii) efficiency, (iv) portability, (v) usability, (vi) testability, and (vii) readability.
- **organisational requirements**: come from strategies and procedures established in the context of the manufacturing process of the system or the client organisation, being examples (i) process standards that must be followed, (ii) implementation requirements, like the programming language to be adopted.
- **external requirements**: have origin in external factors to the system and the development process, being examples (i) interoperability requirements that define how the systems interact with other systems, (ii) legal requirements to guarantee that the system is compliant with the laws, and (iii) ethical requirements to make sure that the society in general will accept the system.

In this book, the classification scheme proposed by Robertson and Robertson (2006), which is especially targeted for the software and information systems domain, is used for structuring the requirements document described in Sect. 7.2. That classification scheme defines eight types of non-functional requirements:

1. **appearance**: the visual aspect and the aesthetics of the system, namely the graphical interface;
2. **usability**: the easiness of utilisation of the system and everything that permits a more friendly user experience;
3. **performance**: aspects of speed, real-time, storage capacity, and execution correction;
4. **operational**: characteristics about what the system must do to work correctly in the environment where it is inserted;
5. **maintenance and support**: attributes that allow the system to be repaired or improved and new functionalities to be anticipated;
6. **security**: issues related to access, confidentiality, protection, and integrity of the data;
7. **cultural and political**: factors related to the stakeholders culture and habits;
8. **legal**: laws, rules, and standards that apply to the system so that it can operate.

There are other classification schemes, some of which are indicated in the "further reading" section of this chapter. Therefore, each project team must adopt a classification reference that is deemed appropriate. In reality, these forms of classification serve only to aid in organising the information. In practical terms, there is no obligation in identifying non-functional requirements of all the types for the very same project. It is not also really very critical to discuss to which type a given non-functional requirement belongs to. The fundamental thing is to recognise that the requirement does exist and to document it.

3.3.1 Appearance

It is common that the client specifies requirements of this type, indicating, for example, corporative elements, logotypes, fonts or colours to be used. If a tailor-made software system is being developed for a given organisation, it is important to take into account its visual identity. In some contexts, the appearance of the system is critical for it to be successful. For instance, sports is a domain where the colours, associated with the clubs, are important elements. The McDonald's restaurant chain was forced to change its traditional colours (red and yellow), when a new restaurant was open in Istanbul (Turkey), near the Beşiktaş stadium. Galatasaray, that uses the same colours of McDonald's, is the rival of Beşiktaş and the fans of this club did not accept in their 'territory' a restaurant with the colours of the rival. The solution was to reconvert all the restaurant, including the big 'M', to black and white (the colours of Beşiktaş), avoiding thus possible problems or boycotts.

> "It is only shallow people who do not judge by appearances."
> *Oscar Wilde (1854–1900), novelist*

As examples of appearance requirements, one can consider the following:

The product shall have a look & feel similar to the remaining products of the company.
The product shall be attractive to teenagers.
The product shall be identifiable with the company that will use it.

3.3.2 Usability

Usability is a critical aspect for the success of many systems. Here, one includes aspects related to the ease of use, personalisation, ease of learning, comprehensibility, and accessibility.

Ease of use is related to the efficiency of utilising a given system and with the mechanisms that exist to reduce the errors made by the users (for example, an automatic spell checker). Here one should take into account the GIGO (garbage in, garbage out) principle, which indicates that software programs that receive incorrect input data, after processing them, generate undesired output data Lidwell et al. (2010, pp. 112–113). The best way to avoid garbage in the output is preventing garbage in the input. For example, when a telephone number must be introduced, it is important to ensure that only digits are typed in and that the number of digits is correct (for instance, in Portugal all phone numbers have 9 digits) and that all possible access codes are available to be chosen (avoiding thus typing in them).

Personalisation is associated with the capacity of adapting the system to the tastes and needs of the users, including the choice of the language, currency, time zone and other configuration options (colours, backgrounds, icons).

Ease of learning is concerned with the way users are trained to use the system. Based on requirements of this type, the development team can prepare training and help procedures for the users. Associated with the ease of learning is comprehensibility that determines if the users intuitively capture the functionalities of the system and how to operate it. In some cases, the system is expected to be so easy to use that no external help is required. Vending machines (for instance, public transportation tickets, parking, cinema tickets) must have very high comprehension levels; if this is not the case, the users resort to other alternatives to satisfy their needs.

Accessibility indicates how easy it is to use the system, for people who are somehow physically or mentally disabled. The disabilities can be related, for example, with physical, visual, auditive, or cognitive aspects.

The following examples are usability requirements:

The product shall be easy to use for illiterate persons.
The product shall be especially intuitive to use for children with 4 years old.
The product shall be usable by visually impaired persons.

3.3.3 Performance

Performance refers to the capacity of a system to respond to its stimulus, that is, the time necessary for responding to the events or the number of events processed by time unit. It is the degree to which a system can accomplish its functionalities, within a given set of constraints. The performance of a system is related to the processing time of the tasks, response time of the operations, accuracy of the results, reliability, availability, fault-tolerance, storage capacity, scalability.

The behaviour of a real-time system must respect, in addition to the intended functionality, a set of timing requirements externally defined. The correction of an answer of the system includes also the instant when it is produced. A late answer is incorrect and constitutes a non-conformity, according to the stipulated behaviour for the system. For example, the time limit to fire an alarm in the case of a dangerous situation must be scrupulously met; otherwise, it may not be possible to save the persons and the goods that are being monitored.

Accuracy of the results is related to the preciseness of the calculations made by the system and how they are stored and shown. These factors are relevant for values related, for example, with hours, GPS coordinates, money (rounding of exchanges bank interests), scientific calculations, and percentages.

Availability quantifies the percentage of time during which a given system is operational and working correctly. It is a measure of the likelihood that a system will be operating when called upon (Kossiakoff and Sweet 2013, p. 258). It is a very important aspect that has ramifications in other issues: confidence of the users in the products that they use, value of the information, processes efficiency, productivity of the organisations. Availability is normally measured by the MTBF (mean time between failures) and MTTR (mean time to repair), i.e., the speed with which the system is able to be available again, after the occurrence of a failure. It is expressed by the following formula:

$$\text{availability} = \frac{\text{MTBF}}{\text{MTBF+MTTR}}$$

For critical systems (either in security or business terms), availability must be very high. This characteristic is not however enough, since the system must also be reliable, since a single error may represent severe damages or even the loss of human lives in more extreme situations. *Reliability* is the capacity of a system to continue in operation over time and is related to the probability of a system producing correct results in a given period of time. Nowadays, there are various businesses that depend totally on computer-based systems that support them (for instance, airline, bank, insurance company, e-commerce, industrial company), being expectable that those systems are "always" available (i.e., that they have short periods of unavailability) and reliable. Reliability is normally measured by the mean time between failures.

Fault tolerance is an important characteristic, since it indicates the capacity of the system in keeping an acceptable level of operation, even in undesirable circumstances. The higher the fault tolerance, the bigger the capacity of the system to continue working, even if some of its components fails.

Data storage capacity specifies the amount of data that the system should be able to process and store. In fact, while the functionality of storing data is a functional requirement, the respective capacity should be seen as a non-functional requirement.

Scalability is the ability of a system to continue to show a high quality of service, even when subjected to a greater number of requests. Scalability can be related with the ability to serve more users simultaneously, treat a higher volume of information, respond to more requests. In any case, it is assumed that this load increase does not imply significant changes to the system in order for it to maintain the performance levels. Scalability of software systems has often to be evaluated taking into account the hardware resources, because it is only possible to determine factors such as response time or number of users that can be served simultaneously, if both the hardware and the software are analysed as a set.

As examples of performance requirements, one can consider the following ones:

The product shall identify the employee based on a photo in less than 3 s.
The product shall operate in local mode, if the connection to the server is lost.
The product shall calculate the interest to the tenths of cents.

3.3.4 Operational

Operational requirements describe aspects of the context in which the system will work. It is relevant to indicate whether or not a system has to be prepared to work, for example, in industrial environments (where there are noise, dust, vibration), contexts of mobility (prolonged absences of network signals, operation with only one hand, low light) or marine environments (high levels of humidity, ripple). The operational requirements also relate to other systems that interact with the system at hand. In these cases, one should indicate how to interoperate with those systems.

As examples of operational requirements, the following may be considered:

The product must continue to function at 30 m under water.
The product shall upload the data in batch from text files.
The product shall export the curriculum vitæ in the Europass format.

3.3.5 Maintenance and Support

System maintenance, in general, is divided into four main types: preventive, corrective, perfective, and adaptive (see Table 2.2). The strategy for maintaining systems is limited by several factors (organisation, context, technology, laws, business), which determines how the different types of maintenance efforts are developed over the time to make sure that the system remains useful and updated. Thus, in systems in which maintenance is paramount, it is convenient to consider maintenance requirements during the analysis phase. *Modifiability* is an attribute strongly related to maintenance and is dependent on how easy it is to locate the system components that must be changed. In principle, it is preferable that the change has impact on a reduced number of components, since it is more costly and difficult to make modifications in the context of many components.

In terms of support, it is important to know, for example, what kind of support and training the users are expected to be provided. It may be necessary to provide technical support services, include help menus in the software system, or make available tutorials that explain how to operate the system. Today, it is also common to make videos that explain how to operate a given product.

The following examples can be considered as maintenance and support requirements:

> The source code of the product programs should contain comments.
> The product must be prepared to be translated to any language.

3.3.6 Security

Security is a measure of the ability of a system to resist unauthorised attempts to access, while continuing to provide its services to authorised users. Security is an aspect of quality that assumes essentially two issues: confidentiality and integrity.

Confidentiality is a set of rules that prevents restricted information from reaching the wrong people and ensures that the authorised persons can obtain information. It refers to the indication of whom is authorised to access the functionalities and data and the circumstances under which the access is granted. A system is said to be confidential if it ensures that only authorised users can manipulate the data to which they have access.

Integrity is related to the reliability and accuracy of the information. It involves keeping the consistency and accuracy of the data throughout the system life cycle. Integrity specifies the mechanisms that exist to prevent data to disappear or to be damaged, in case of unwanted events (e.g., electromagnetic phenomena, equipment failures, fires, floods, computer attacks) or misuse of the system. For example, the existence of a backups policy or automatic mechanisms to store information are forms of tackling the unwanted events that may affect the data.

As examples of safety requirements, one may consider the following:

> The information related to the performance evaluation of an employee shall be provided only to that employee and their superiors.
> The product shall ensure that only registered users have access to the clinical data of the patients.
> The product shall reject the introduction of incorrect data.

This last requirement can also be seen as a requirement to ease the use of the system, since it reduces the occurrence of errors made by the users.

3.3.7 Cultural and Political

Culture can be seen as the characteristics of a particular group of people, defined by everything from language, religion, cuisine, social habits, behaviour patterns, music, and arts. Cultural requirements are especially critical when a product is commercialised in different countries. They are also relevant if the product serves different professional groups, due to the differentiated cultures that exist from profession to profession. For example, persons connected to artistic fields (painting, music, architecture, theatre) have in general, more sensitivity to the aesthetic issues of the product than those linked to more technical areas (engineering, science, management). Issues of this type that are applicable to the product that is under development must be specified as cultural requirements.

Political requirements are specific to the factors related to the strategy and the powers in the organisations that affect the system development. These requirements have sometimes a difficult rationale. A company may wish to incorporate in a given system components of a specific supplier, only based on a given strategy or the internal interests. Often, a requirement of this type must be accepted, even if it is possible to find more efficient or cheaper alternatives.

As examples of cultural and political requirements, consider the following:

The product shall use Canadian English.

The product shall show the local holidays in the calendar.

The product shall use components manufactured in the European Union.

3.3.8 Legal

Any system, regardless of the technology, has obviously to respect the established laws. Thus, development teams should consult lawyers, legal advisors and jurists, for knowing whether any law or rule is being broken by the system. For example, in Portugal, laws recently introduced (e.g., ordinances 363/2010, issued on June 23rd, and 22-A/2012, issued on January 24th) regulate the procedure for certification of software products that deal with invoices, by defining a set of technical rules to be met by the software producers. The current EU Data Protection Directive 95/46/EC does not take into account relevant issues like globalisation and technological developments (for example, social networks and cloud computing). This situation has led the European Commission to work towards unifying data protection across 28 countries within the European Union, by proposing a single law, the General Data Protection Regulation (GDPR).

"In 2031, lawyers will be commonly a part of most development teams."
Grady Booch (1955–), software engineer

The following requirements belong to the legal type:

The product shall be aligned with the PMBoK guide.
The product shall be certified by the Taxing Authority.
The product shall fulfil with the copyright.

3.4 User and System Requirements

In spite of the many possible definitions for requirement, the one that is adopted in this book portrays reasonably well the differences on perspective between the users and the clients, by one hand, and the engineers, by another hand. It can give origin to two types of approaches for capturing requirements.

A **user requirement** represents a functionality that the system is expected to provide to its users or a restriction that is applicable to the operation of that system. These requirements are strongly related to the problem domain and are normally expressed without great mathematical rigour, using natural language and informal diagrams. This permits stakeholders to read, analyse, and discuss those requirements. According to the discussion presented in Sect. 5.2, a more adequate designation for this type of requirement would be *stakeholder requirement*, since one must address the necessities of all stakeholders and not just those of the users. However, the expression 'user requirement' is so widespread that it becomes counterproductive to replace it by another one, even if more appropriate from a terminological point of view.

A **system requirement** constitutes a more detailed specification of a requirement, being generally a desirably formal model of the system. These requirements are oriented towards the solution domain, providing informations to the engineers to help them in the system design and construction. They are thus requirements that are documented in a more technical language than the one adopted for the user requirements. System requirements are an intermediary stage between user requirements and system design. Nevertheless, it is desirable that they are independent from design and implementation pre-decisions.

The dichotomy between these two types of requirements implies different approaches, either in relation to the proper level of detail for each type of requirement, or with respect to the several available notations to represent them (Illustration 3.2). The next example presents one user requirement and four system requirements that can be derived from it.

Illustration 3.2 User and system requirements are relevant in different phases of the development process, being important to preserve the relationships between them

---

user requirement

A user manipulates files created by himself or other users.

system requirements

1. The file types and the respective icons are defined by the users.
2. Each file type is represented by a distinctive icon.
3. Each file type is associated with a program that processes and manipulates the corresponding files.
4. When a user clicks on a file icon, that file is automatically open by the associated program.

The requirements result from necessities that exist in the problem domain to be addressed. Consequently, user requirements should be described with the problem domain terminology. If this recommendation is followed, it is more likely that the requirements can be focused on the questions associated with the problem that is being addressed. Engineers should avoid to speak with the terminology of the technological domain of the system under development. If one proceeds this way, the premature inclusion of solution domain issues is avoided.

> "If you talk to a man in a language he understands, that goes to his head. If you talk to him in his language, that goes to his heart."
>
> *Nelson Mandela (1918–2013), politician*

This focus on the problem domain, rather than on the solution domain, seems evident, but it is common to fall into the temptation of looking to the challenges of a project in a technological perspective. In these cases, the tendency is to defocus from the true problem and to concentrate the attention on the alleged solution. As everyone knows, it is only possible to develop an adequate solution, if the problem to be solved is well characterised. Unfortunately, it is not unusual for teams to have the tendency to develop solutions for problems that are poorly formulated.

The focus on the problem, instead on the solution, allows in some cases, unquestionable gains. In a given company, the managers were interested in being sure that the number of meals to be paid to the service provider was exactly the one that was effectively served. Hence, the client requested a solution that implied the use of transponders[1] that would be put in the identification cards of the employees, so that one could count how many took their lunch at the canteen. In that solution, whenever an employee would go through the door, where a transponder detector would be installed, the respective lunch would be summed up. The client clearly described what he thought as "the" solution, but was not focused on the problem that would have to be solved. Later, with a focalisation on the problem, it was possible to realise that the transponder-based solution was more expensive than the one that eventually was adopted. The built solution was based on using directly the attendance management system already in operation at the company and requesting the employees to indicate if they would (or would not) have lunch at the canteen. Thus, the solution was simply making available the list of the employees that were to have lunch at the canteen.

3.5 Related Concepts

There are other terms, used by the software community, that are conceptually related to 'requirement', yet with a distinct meaning. This set includes the following words: necessity, functionality, feature, and service. For each one, a small discussion about its scope is presented, to help understand the difference and similarities among them.

In a psychological perspective, a necessity is the perception of the lack of something that is considered indispensable, useful or convenient. A **necessity** is the consequence of a problem that a person or an organisation has and that must be addressed to justify the usefulness of the solution that satisfies it. The necessities represent difficulties, opportunities, or expectations associated with the problem domain, thus providing an indication about what the solution should solve.

The necessities can be structured according to a pyramid, based on the hierarchy of needs theory, presented by Maslow (1943, 1954). This theory determines that the lower level necessities must be satisfied before the higher level ones. According to Lidwell et al. (2010, pp. 124–125) the five principal levels of necessity are: functionality, reliability, usability, proficiency, and creativity.

---

[1] A transponder is a device that emits an identifying signal in response to an interrogating received signal.

"Necessity is the mother of invention."

proverb

Another term frequently used is **functionality**, which is defined as a capacity of a given system to provide a useful function. In practice, a functionality corresponds to the concretisation in the final solution of what was identified as a functional requirement in the analysis phase.

A service derives from the relationship that is established between a system and its users, with the purpose of providing a solution to a given problem. A **service** aggregates a coherent set of related functionalities, that allow users to satisfy a given necessity or objective. Hence, a service represents a high-level functionality. For example, the services provided by an automobile include the transportation of persons and goods by road, according to the safety rules established by the law. This transportation service implies a cohesive set of functionalities, such as accelerating, breaking, and signalling the change of direction.

The term *feature* has several meanings, which depend on the domain and context in which it is used (Classen et al. 2008). In this book, a **feature** is a characteristic of a product or system clearly perceived as distinctive or differentiating by the users and that cohesively aggregates a set of functional and non-functional requirements. For example, the indication that an automobile must be convertible corresponds to the indication of a *feature*, that has many implications for the design and the manufacturing of the automobile, such as functional, aerodynamical and safety-oriented questions.

3.6 Summary

Requirements represent the necessities of the users and the constraints that are applied to a system and that must be considered throughout the development. The requirements can be classified, according to a first criterion, as either functional or non-functional. While the former represent the functionalities available to the users of the system, characterising its behaviour as an answer to the stimulus that it is subject to, the latter represent the quality attributes associated with the system to be developed. The non-functional requirements are divided in this book in eight different types: appearance, usability, performance, operational, maintenance and support, security, cultural and political, and legal. Non-functional requirements are extremely important, because it is common to initiate many projects due to the non-functional aspects. If the current system is expensive, very unreliable, unsafe, or difficult to maintain, that often conducts to the need of creating a better system, in which the non-functional requirements have a very high priority.

According to a second criterion, the requirements can be designated as either user or system requirements. User requirements represent functionalities that the system must offer to its users or restrictions that are applicable to the operation of that system.

These requirements, strongly related to the problem domain, are usually expressed in a natural language. A system requirement is oriented towards the solution domain and is a detailed specification of a requirement, generally in the form of a formal model of the system. This chapter also relates the term 'requirement' with other terms (necessity, functionality, service, and *feature*), discussing the respective scope with the purpose of clarifying the differences and resemblances among them.

Further Reading

A characterisation about what is a requirement can be found in any book about requirements engineering or even software engineering. Sommerville (2010, Chap. 4) discusses the various types of requirements and presents illustrative examples. Robertson and Robertson (2006, Chaps. 4 and 5) discusses functional and non-functional requirements and also present various examples of those types of requirements. Pohl (2010, Chap. 2) defines the term 'requirement' and characterises the different types of requirements. Chemuturi (2012, Chap. 2) provides an interesting classification of requirements, based for example on functionality considerations, product construction considerations, or source of requirements.

In addition to the two classification schemes for functional requirements discussed in the chapter, there are other alternatives. The ISO/IEC 9126 standard is focused on quality, proposing quality attributes, divided in six main characteristics, for evaluating software products: functionality, reliability, usability, efficiency, maintainability, and portability. There are equally other proposals suggested by several authors, for example, Boehm et al. (1978); Roman (1985); Grady and Caswell (1987); Jureta et al. (2006); Meyer (2013), which demonstrates the lack of a consensus in the software engineering community with respect to this question.

The area of healthcare is heavily regulated, so the development of eHealth (healthcare practice supported by electronic processes and communication) must account for conventional health law. Purtova et al. (2015) presents an integrated approach to legal requirements engineering in the context of eHealth, bringing together a methodology for mapping existing legal and regulatory landscape and the strategies to interface the identified rules into design of the eHealth technology and processes. Bernsen and Dybkjær (2009) address how to develop and evaluate multimodal systems which are usable by people.

Exercises

Exercise 3.1 (Naveda and Seidman 2006, pp. 39–40) Which is the type of elements appropriate to be included in a requirements document?

1. design restrictions
2. product delivery constraints

3. functionalities to make available
4. performance characteristics

Exercise 3.2 (Naveda and Seidman 2006, pp. 33–34) Which is the type of requirements that should be included in a requirements document?

1. functional requirements
2. maintenance requirements
3. project requirements
4. performance requirements

Exercise 3.3 (Naveda and Seidman 2006, pp. 41–42) Which is the element that must be included in a requirements document?

1. acceptance/validation procedures
2. delivery plans
3. quality attributes
4. activities to guarantee the quality

Exercise 3.4 (Naveda and Seidman 2006, pp. 57–58) Which of the following arguments is the most solid/strong to justify the specification of the non-functional requirements of a system?

1. The non-functional requirements should only be considered in development contexts subject to tight restrictions (resources, budget, or deadlines).
2. The non-functional requirements are only external characteristics of the system and can be obtained later.
3. If a functionality is present in the system, the non-functional requirements determine how usable and useful it is.
4. The non-functional requirements take less time to specify than the functional requirements.

Exercise 3.5 Consider the following requirement:

The system should be easy to use for trained persons.

---

1. Classify this requirement with respect to its type.
2. Is this requirement verifiable? Justify.
3. Rewrite the requirement so that it becomes measurable.

Exercise 3.6: Ghezzi et al. (1991, pp. 18–36) indicate the following qualities as being the most important for software products and processes: (1) correction, (2) reliability, (3) robustness, (4) performance, (5) facility of utilisation, (6) verifiability, (7) maintainability, (8) reparability, (9) evolvability, (10) reusability, (11) portability, (12) comprehensibility, (13) interoperability, (14) productivity, (15) actuality, (16) visibility.

1. Define each one of those 15 qualities in a succinct but rigorous way.
2. Classify each quality according to the three categories of non-functional requirements suggested by Sommerville (2010).
3. Classify each quality according to the eight types of non-functional requirements proposed by Robertson and Robertson (2006).
4. Bass et al. (1998, p. 76) divide the quality attributes of a software systems in two groups: (1) those that are observable in execution time and (2) those that are not observable in execution time. Based on that division, classify the 15 qualities among those two groups.

Exercise 3.7: Classify the following non-functional requirements according to the classification scheme proposed by Robertson and Robertson (2006):

R1: The product shall be easy to use for persons that do not dominate the English language.

R2: The product shall follow the new orthographic agreement for the Portuguese language.

R3: The product shall be available 24/7/365.

R4: The product shall be able to register 50,000 items per hour.

R5: The product shall present the money values in the currency chosen by the users.

R6: The product shall fit in a trouser pocket.

Exercise 3.8: The following text was obtained from a client that considers it as a good description of what is needed:

(a) The project consists in a database to store historical information about the performance of the athletes of a given club. (b) Users will be able to know which were the original training sessions and (c) to compare them with the sessions actually performed. (d) The system will be easy to use to all coaches and (e) it will be accessible from all the workstations in the club. (f) The system will be developed in Java and (g) must have an acceptable reliability. (h) The system will produce at the end of each month updated information about all the athletes for the main coach. (i) The screen will show tables with the planned sessions versus. the sessions that were not initiated or completed, and (j) will estimate the date for concluding the project. (k) The system must be operating satisfactorily by December 2019 and (l) will be produced by the IT department, (m) under coordination of the engineering department. (n) The software must run in PCs and smartphones.

Classify the sentences (a) to (n) according to the following types: (i) User requirements, (ii) System requirements, (iii) Design elements, (iv) Plans, (v) Context information, (vi) Irrelevant details.

Chapter 4
Requirements Engineering

Abstract The chapter begins with a general discussion about the requirements engineering area, with the aim of introducing what is its scope of activity and what are its purposes. Next, the various activities that make up the requirements engineering process are presented, emphasising those activities that are deemed to be fundamental. The chapter closes with a debate about the main challenges and problems which the requirements engineers are faced with.

4.1 Definition of Requirements Engineering

It is convenient to discuss here the origin and the meaning of the term that provides this chapter its name. The 'requirements engineering' term is relatively new and was introduced by the scientific community to designate all the activities related to requirements discovery, negotiation, documentation, and maintenance in engineering projects. As indicated in Sect. 2.4, it was (and still is) common to employ the term 'analysis' (or variants like 'requirements analysis' and 'systems analysis') to refer to this set of activities, but, in this book, the new designation is preferred.

Requirements engineering is inherently broad, interdisciplinary and is constantly open. It is related to the transformation of informal descriptions of the real world into specifications in languages with a rigorous and mathematical basis. For this reason, it may seem more obscure than other engineering approaches, like, for example, the testing engineering and the maintenance engineering. This perception comes probably from the fact that the sociologic dimension is much more crucial for requirements than for other engineering areas (Goguen and Linde 1993). The use of the word 'engineering' is premeditated and with the clear intention of highlighting the need to utilise systematic techniques and methods to make sure that the requirements are clear, complete, coherent, relevant, and measurable. Moreover, engineering, generically speaking, is not, contrarily to what is considered by many, a purely technical discipline, based on precise calculations and complex and already established concepts. Instead, the engineer is a creative professional and her acts depend on a well-balanced mixture of science, technique, art, experience, and common sense. Therefore, the use of the word 'engineering' is perfectly adequate and justified in the scope of requirements.

© Springer International Publishing Switzerland 2016   65
J.M. Fernandes and R.J. Machado, *Requirements in Engineering Projects*,
Lecture Notes in Management and Industrial Engineering,
DOI 10.1007/978-3-319-18597-2_4

"Common sense is the most widely shared commodity in the world, for every man is convinced that he is well supplied with it."

René Descartes (1596–1650), philosopher

There are several definitions for requirements engineering proposed by different authors, and so some of the most relevant ones are here discussed. In a simple way, requirements engineering consists in the study of a problem that leads to the system development, before taking any design or implementation action.

For Zave (1997), requirements engineering, in the scope of software engineering, is focused on the real-world objectives established for the functionalities and the restrictions of software systems. It is also related to the relation of those factors to precise specifications of the software behaviour and to the evolution of those specifications.

Sommerville (2010, p. 83) mentions that requirements engineering is the designation given to the process of discovery, analysis, documentation and checking the services and restrictions related to the operation and development of software systems. It is a set of structured activities that, with respect to a system, aid in obtaining an understanding about the domain, the restrictions of operation, the functionalities requested by the stakeholders and the essential characteristics. Requirements engineering addresses also the system documentation, to be used by the stakeholders and the engineers that participate in its development.

However, one can have a wider perspective, viewing requirements engineering in the scope of systems engineering. For Hull et al. (2011, p. 8), requirements engineering is concerned with the discovery, development, traceability, analysis, testing, communication, and management of requirements with the objective of defining the system at different levels of abstraction. Requirements engineering is focused also on the partition of the system under development in various parts, defining which requirements must be assigned to each parts. It is recommendable to execute these tasks at an adequate abstraction level, so that the technology is not exploited in detail, since often it is not possible or desirable to define from the very beginning the technological issues related to the blocks.

According to the perspective supported by van Lamsweerde (2000), requirements engineering handles the identification of the objectives to be achieved by the system at hand, the operationalisation of those objectives in services and restrictions, and the assignment to agents (persons, existing devices and technological systems to be developed) of the responsibilities resulting from the requirements.

Based on these various definitions, **requirements engineering** is seen in this book as the set of activities that, in the context of the development of a system through an engineering project, permits eliciting, negotiating, and documenting the functionalities and the restrictions of that system. This definition is sufficiently generic to be applicable to systems in all engineering domains, which frames requirements engineering in the systems engineering sphere. The final result of requirements engineering consists of a set of artefacts (documents, models, specifications) that permit one to comprehend the relevant aspects of the system and to proceed with its manipulation (for instance, verification, simulation, and animation), so that one can predict the effect of possible changes. Those artefacts facilitate communication with the stakeholders, through the ideas expressed on them.

Requirements engineering, as a discipline, includes the body of knowledge that aims to help the development teams to better understand the problem that has to be faced, obtaining the description of the requirements for the system that is developed to solve that problem. The objective of requirements engineering is to increase the chances of the system under development to satisfy the future users and to answer in an acceptable way to their necessities. In summary, requirements engineering is a process that, according to Pohl (2010, p. 48), seeks to ensure the three following objectives:

- all the relevant requirements are explicitly known and comprehended at the intended level of detail;
- a reasonable and wide agreement about the requirements is obtained among the stakeholders;
- all the requirements are duly documented, in conformity with the established formats and templates.

To achieve those objectives, requirements engineering addresses the study of the problem domain, conducting to a description of the system, seen as a black box, indicating the intended functionality and the operational characteristics (for instance, reliability, availability, usability, and performance), but ignoring the aspects of the internal structure that allow that behaviour to materialise.

Requirements engineering determines what the system must do to meet the necessities of users and not how it should be built. So, analysing a system consists in studying it without taking into consideration issues related to the implementation technology. In requirements engineering, one dedicates an appreciable effort to systematise the knowledge of the problem domain. Requirements engineers, in conjunction with the various stakeholders, should identify what the system will do. To achieve this aim, the engineers resort to questionnaires, interviews, observation of the activities, manuals, reports, and other documents to collect the information they need. It is usually difficult, although totally desirable, keeping the requirements strictly separated from their own solutions. It is expected that the requirements of a given system are necessary, clear, correct, complete, viable, traceable, verifiable and negotiable.

One of the main objectives of requirements engineering consists in establishing communication channels between the holders of the problem (i.e., the users and the clients) and those that will construct a solution (i.e., the systems engineers). The requirements must be managed by requirements engineers, jointly with representatives of the client, key users and other domain specialists.

The quality of the work performed by the requirements engineer has a decisive impact on the remaining development phases. It is crucial that it does not contain defects (i.e., decisions that are likely to be altered), since, as they propagate to the next phases, they imply expensive corrections, as Fig. 4.1 illustrates. In the case of the software domain, Robertson and Robertson (2006, p. 262) indicate that more than

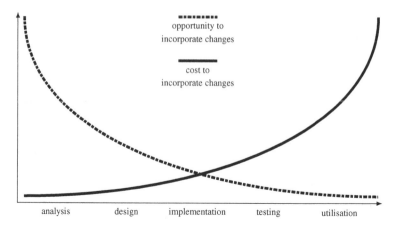

analysis   design   implementation   testing   utilisation

Fig. 4.1 Opportunity and cost of incorporating changes during the system development process

half of the defects made during the development are introduced in the analysis phase. Even if the order of magnitude of that percentage could be questioned, especially in other engineering branches and fields, it seems that there are no doubts with respect to the impact that the quality of requirements engineering tasks has on the success of any development project. If a requirement is misunderstood, it is probable that the design decisions based on it are inadequate; similarly implementation and testing could lie in incorrect assumptions. Hence, it is convenient to detect as soon as possible the changes to be incorporated, since the associated costs to its concretisation increase as the development advances, as shown in Fig. 4.1.

> "It is easier to confess a defect than to claim a quality."
>
> *Max Beerbohm (1872–1956), parodist*

4.2 Activities

Requirements engineering is not a mere set of activities isolated from the rest of the development process, but it requires to be permanently revisited, which implies the need of several closed feedback loops (Aurum and Wohlin 2005b). The requirements engineering process, according to Pressman (2009, p. 121), include seven principal activities, as depicted in Fig. 4.2: (1) inception, (2) elicitation, (3) elaboration, (4) negotiation, (5) documentation, (6) validation, and (7) management.

The sequential character shown in Fig. 4.2 must be seen with some relativity. The processes in engineering contexts, excluding some very specific exceptions, are not executed in that precise order, neither are chronologically so well delimited. In reality, the processes tend to be collaborative, cooperative, iterative, and incremental. By one hand, the cooperation and collaboration amongst the requirements engineers

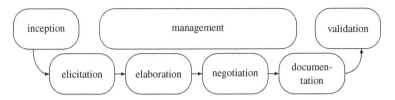

Fig. 4.2 Main activities of the requirements engineering process

and between them and the stakeholders presupposes parallelism/concurrency in the execution of the tasks. By another hand, the iterative and incremental nature of the process means that a task previously executed can be executed again, as many time as needed. These two realities dismantle without doubts the linear character of the figure. In practice, it is possible to move from any activity to another one, either forwards or backwards, implying that the linearity shown in the figure must be understood as a simplified and idealised process flow at a very abstract level. This proviso applies equally to all the other processes presented in this book; even to the waterfall process, as discussed in Sect. 2.4.1. Despite these degrees of freedom, it is expectable that some of the ideas, principles, and characteristics of the processes are to be followed.

Although all the activities of the requirements engineering process are obviously important, there are three of them that are considered as fundamental to its pursuit: elicitation, negotiation, and documentation (Illustration 4.1). Pohl (2010, pp. 48–50)

Illustration 4.1 The requirements engineering process involves three main activities and produces outputs that are used in the subsequent phases of the development process

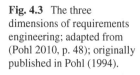

Fig. 4.3 The three dimensions of requirements engineering; adapted from (Pohl 2010, p. 48); originally published in Pohl (1994).

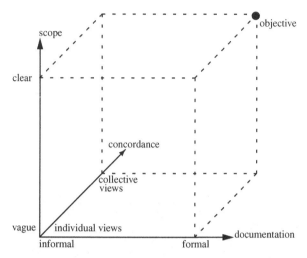

explains that these three activities contribute directly to the three dimensions that must be fed during the requirements engineering process: scope, concordance, and documentation[1] (Fig. 4.3). Thus, the requirements elicitation activity contributes to the progress of the 'scope' dimension (initially vague and at the end clear). The negotiation activity feeds the 'concordance' dimension, through the identification and resolution of conflicts. Finally, the documentation activity is associated with advances in the dimension with the same name, making the initially informal documents into documents compliant with the adopted rules and formats. These activities are those that deserve specific chapters: requirements elicitation (Chap. 5), negotiation (Chap. 6) and documentation (Chaps. 7 and 8). These activities were also the ones mentioned in the definition proposed for requirements engineering.

Inception

Any development project requires some *inception* mechanism. Someone must initiate the process, based on some necessity or business expectation, since that no engineering project starts spontaneously. Without that formal trigger, there will be no project. This activity has a vital importance for the society in general, since the need that was identified appears with the generic perspective of enhancing the life quality and creating value.

This initial project step lies invariably in the identification of a necessity. It is common that the perception of that necessity occurs due to the dissatisfaction in relation to same aspect of the current situation. In this scenario, the development team is habitually contacted by someone to develop (or improve) a given system. The way in which the contact is done varies a lot from case to case; for example, a meeting can be requested by the client to explain her necessities. The inception can have its origin in the study of the problem documentation that details among others things

---

[1] Please notice that the 'documentation' term is used to refer to both the activity and the dimension.

the intended requirements. This documentation can vary, from project to project, in its form, size, and contents. Thus, the development teams must be prepared for any type of document (or even in a more extreme situation to its nonexistence) and, independently of the contents or format, use it to accomplish its task.

In the cases in which the client is internal to the system producer, the inception is requested by the persons (for instance, business managers, marketing specialists, or product managers) that interact more directly with the users. Such experience has allowed them to identify the necessities that are not yet satisfied and that were evaluated as having potential to bring financial profits to the organisation. In this case, those persons act as users' representatives.

During this activity, the development team collects the first informations about the system, to study the project feasibility. This recollection must be done in width and not in depth, that is, the objective is to understand the full system scope and not the details. In this activity, one must obtain the stakeholders concordance with respect to the project scope, objectives and plan. At the end of the activity, the requirements engineer should be able to describe what is the client vision and return on investment, that is, what the system is intended to do, what is behind the client need and why he is willing to spend money in the development of that system. One must also evaluate, in this step, if what the client needs is already available in the market and ready to be used.

> "The answer to a feasibility study is almost always 'yes'."
> *Robert L. Glass (1932–), software engineer*

There are groups of questions that can be asked during the execution of this activity, in order to focus the stakeholders on the issues that must be effectively addressed. Next, some questions for that purpose, suggested by Gause and Weinberg (1989) and adopted by Pressman (2009, pp. 127–128), are presented. These questions are divided in three groups. In the first phase, it is important to assess what are the main system objectives and the benefits that will result from its utilisation in real contexts. The questions that can be asked for that purpose are:

- who is behind the request for this system?
- who will use it?
- what are the economic benefits of the system?
- are there other sources that is necessary to use/consult?

In a second phase, it is important that the requirements engineer acquires a better understanding of the problem and that the client can verbalise his vision about the system:

- what characteristics do you wish to see included in the outputs generated by the system?
- what problems are solved by the solution?
- can you describe the business environment in which the system will work?
- will the system be affected by any performance issue or restriction?

In the third (and last) phase of questions, the focus lies in understanding if the communication between the parts has worked as expected. The recommended questions are:

- do you consider yourself the right person to answer my questions?
- do you think that my questions are pertinent to the problem that you have?
- am I asking you an excessive number of questions?
- are there questions that I was supposed to ask you?
- is there someone else that can provide additional informations?

> "My greatest strength as a consultant is to be ignorant and ask a few questions."
> *Peter F. Drucker (1909–2005), management consultant*

It is not expectable that this activity takes too long, especially because, due to the nonexistence of a contract between the client and the system provider, this activity constitutes an investment with the prospect that it can materialise into a project. As a result of this activity, one should produce a document where some predictable system development scenarios are anticipated. The outputs for this activity are: (1) the characterisation of the problem, (2) the definition of the alternative solutions and the respective benefits, and (3) the identification of the resources, costs, and dates for the proposed alternatives.

Requirements elicitation

This activity handles how requirements should be captured. The requirements elicitation techniques must identify the sources of requirements and should also aid the various stakeholders to correctly describe the requirements (Goguen and Linde 1993). This activity is inherently communicational, since it requires an in-depth interaction with the stakeholders. Hence, it is impossible to automate the essential aspects of this activity. Interview, survey, introspection, ethnography, focus group, cooperative work, domain analysis, object-orientation, prototyping, scenario, goal modelling, and persona are important requirements elicitation techniques that Chap. 5 describes in detail.

> "The most effective techniques is speaking to real potential users, watching real potential users doing their current job, and allowing users to play with experimental early versions of the product, commonly called prototypes."
>
> *Alan M. Davis (1949–), requirements engineer*

Requirements elaboration

This activity, sometimes referred to as requirements reconciliation, aims to analyse and classify the elicited, but not yet handled, requirements. Laplante (2013, p. 12) lists various problems that can require the analyst intervention, especially whenever the requirements:

- do not make sense;
- are in contradiction among them;
- are incoherent;
- are incomplete;
- are vague.

It is usual in this activity to organise the requirements in cohesive groups. One of the criteria that is normally used to execute this grouping aims to facilitate the construction of the first system architecture, where one can identify the main components (or subsystems). This situation is one more proof that the development processes do not have a sequential nature and that, in reality, their activities are interrelated in many different ways. With that architecture, it is possible to associate requirements to each component, which aids into the transition for the subsequent development process phases. In this activity, one defines the system frontier and the form which it interacts with the environment, through a context diagram (or through a use case diagram; see Sect. 8.4.2).

Requirements negotiation

Requirements engineering is, in its essence, an intricate process of communication and negotiation that implicates various stakeholders. They are responsible for deciding what to do, when to do, what type of information is necessary, and what tools must be used. In many cases, it is inevitable that conflict situations arise among the considered requirements, which implies the need to promote negotiation mechanisms among the stakeholders. Negotiation in general consists in a collective search for regulating the divergencies and its result can have a significative impact on the acceptance of the final system. In reality, this activity is conducted in parallel with the activities mentioned before and continues until implementing the requirements. Another form of handling conflicts consists in adopting prioritisation techniques, to sustain the choice of the requirements subset to be implemented at each instant. Chapter 6 discusses some relevant issues related to this activity.

Requirements documentation

The term 'documentation' can be used to refer to the activity of documenting, but also to refer to its result, normally a set of documents. Requirements documents are important since they serve as the principal reference to the subsequent phases of the development process. In this activity, to ensure a good structure, quality, and verifiability, the requirements document is normally organised according to two distinct perspectives:

1. user requirements, that describe the expectations and the necessities of the users;
2. system requirements, that establish the agreement between the client and the development team.

The structure and the formality level of the documentation should vary in line with the system characteristics and the nature of the adopted process to develop it. In some cases, a document totally written in natural language can be sufficient, while in others the utilisation of diagrammatic specifications can be mandatory. Chapters 7 and 8 present issues related to the documentation, namely writing requirements in a natural language and graphical and visual modelling.

Requirements validation

The objective of this activity is to ensure that the requirements define the system desired by the client. For such objective, one should examine the requirements document through inspections or technical reviews of the specifications, in order to evaluate if it describes the intended system. This activity is very relevant since the propagation of the defects in the requirements to the next development process phases usually involves alterations at very high costs.

It is discussable if this activity belongs to requirements engineering or to testing. According to a philosophical evaluation, validation is a testing activity, since taking into account the definition made in Sect. 1.4, it covers all the development lifecycle. However, the important message to retain here is that, while the requirements engineering activities are conducted, it is necessary to execute tasks that allow requirements to be verified and validated.

Requirements management

Throughout all the system lifecycle, the requirements set is constantly changing, as discussed in Sect. 4.3. Mechanisms to manage that instability context are needed, in order to evaluate the impact that the changes in the requirements can have on the project. Not all change requests must be accepted. In principle, one must reject changes that imply a significative increase in cost, a postponement of the final delivery, or a system devaluation for the user. The requirements management activity seeks to aid the development team to identify (to assign a unique identifier), control and trace the requirements and their changes. This activity supports all the other requirements process activities (and even the full development process), being executed in parallel with those activities, as illustrated in Fig. 4.2. Requirements management must be executed even after the conclusion of the first version of the requirements document, since the changes in the requirements appear throughout the development process.

In the requirements management context, the construction of several traceability tables permits the development team to relate the requirements with many of their aspects. Requirements **traceability** permits recording and following the requirements lifecycle, both upstream and downstream (Loucopoulos and Karakostas1995, p. 85). Irrespective of the moment in the system lifecycle in which a change request appears, *impact analysis* is a management mechanism that aims to estimate the consequences of those changes. The cost calculation associated with the changes must be used to decide if they are or not materialised, based, for example, in a cost-benefit analysis.

"You can't just ask customers what they want and then try to give that to them. By the time you get it built, they'll want something new."
Steve Jobs (1955–2011), cofounder of Apple

4.3 Challenges and Problems

Generically, the challenges faced by the professionals embraced in requirements engineering are distinct from those which is confronted the remaining engineering community. The requirements are mainly related to the problem domain, while the other artefacts resulting from the technological development activities can be typically placed in the solution domain. Ideally, the descriptions of the requirements are expressed with problem domain terms and concepts and specify how the problem domain is affected by the system to be built. The technological artefacts related to the subsequent process phases are focused on the internal elements of the system. In a simple way, one can say that requirements engineering tries to characterise with rigour the problem in hand, while the other process activities define and refine a given solution for the system to be developed. This distinction is due to the fact that the requirements engineering activities are intrinsically hard to execute, with several factors contributing to it.

Communication problems between requirements engineers and users are common (Illustration 4.2). Users have generally some difficulties in expressing their real necessities, in a perceptible and precise way. Moreover, in many situations, the users only have a vague idea of what they really want. Users and engineers have very different perspectives with respect to the nature of the problem, so they also have different understandings. Even in the cases in which the users have a total awareness of their necessities, something that seldom happens, it is not simple to transfer or verbalise that information. The most common forms of transmitting that information is orally and through written documents, complemented if appropriate with some graphics.

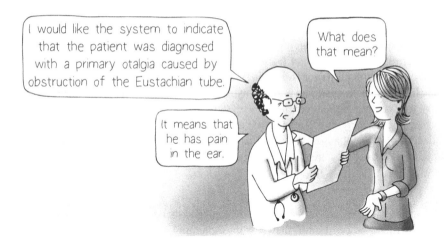

Illustration 4.2 The use of professional jargon hinders understanding

These documents are hard to understand, due to the ambiguity present in natural languages (see Sect. 7.3), so they are susceptible of being interpreted differently, from person to person. The requirements engineers are initially faced with ideas poorly elaborated and often contradictory about what the system must do, but still it is expected that they are able to come up with a complete, coherent, and detailed specification. Table 4.1 summarises some communication difficulties between engineers and clients and suggests possible solutions to solve them. In the specific case of software, one can say that nowadays the major development challenges are essentially related to the communication and not so much with technical issues.

Table 4.1 Communication difficulties among the engineers and the clients

Difficulty	Solution
The client is not able to verbalise what he desires	To observe the users performing their functions in the real context
The client did not notice that he explained incorrectly the problem until he receives a solution that does not satisfactorily solve it	To make sure that, prior to start the system development, the problem to be handled is well formulated and corresponds to the reality
The engineer thinks he knows more about the problem of the client than the client himself	Make the engineer feel the difficulties faced by the users in the real context
Each stakeholder naively believes that all other stakeholders are motived and with a good faith	Identify those that can be against the system development and find ways to motivate them to contribute positively

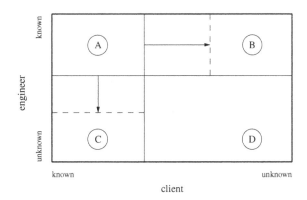

Fig. 4.4 Variant of the Johari window, applied to the known and unknown requirements, both to the client and the engineer

"Software engineering is a field in which members of one culture create artefacts on behalf of members of another culture."

Hans van Vliet (1949–), software engineer

To better evaluate the communication difficulties amongst clients (those that want to see the problem solved) and engineers (those that will develop a solution for that problem), please consider Fig. 4.4. That figure shows a matrix, based on the Johari window (Luft 1969), that divides the space in four quadrants. The window crosses both the known and unknown requirements, either to the client or to the engineer. Each axis should be seen as a theoretical *continuum*, from the total lack of knowledge up to the total knowledge. In reality, there are more than two parties engaged in the process and there are not just engineers and clients. This matrix constitutes obviously a very simplified view of the reality, but serves the purposes of the discussion taken here.

At the beginning of the project, the client knows some requirements that he wants to see incorporated in the system (left hand side of the matrix, composed of quadrants A and C), but, due to many reasons, there are other requirements that he is not aware of (right hand side of the matrix, composed of quadrants B and D). Similarly, the engineer initially knows some requirements, namely the implicit requirements, that will be incorporated in the system (top part of the matrix, composed by quadrants A and B). But, it is obvious that the engineer does not know many of the requirements that the system will incorporate (bottom part of the matrix, composed by quadrants C and D), namely those that are related to the specific needs of the client and that were not yet verbalised, since the project is still at the beginning. Quadrant A represents the requirements that both parties know that will be incorporated. The area of this quadrant, at the beginning of the projects, varies from case to case, but is bigger when the client knows very well the problem to be considered and the engineer has an extensive experience in the domain.

The client and the engineer, during the project, communicate among them, to arrive at a mutual understanding about the requirements to be incorporated in the system. This communication process aims to increase the knowledge of both parties about the requirements. Throughout the communication production that they establish, the engineer gets to know better what the client indeed desires. This modification is represented in the figure by the expansion of quadrant A to the bottom and by the consequent diminution of quadrant C. Additionally, the client increases his knowledge about the system requirements, based on informations that the engineer has shared (and that the system can include). This modification is represented in the figure by the expansion of quadrant A to the right and by the consequent diminution of quadrant B. These two modifications only occur if the client and the engineer share the knowledge that they possess.

The requirements located in quadrant A are relatively simple to capture. They are expected to be addressed during the project, since the two parties know them. The requirements in quadrant B are more arduous, since it is required that the client is persuaded by the engineer about the relevance of those requirements. Here, it may happen that the engineer decides to include those requirements in the system, without consulting the client, since he considers those requirements so evident that they should not be communicated and negotiated. The requirements in quadrant C are even more problematic, since the client is not always able to verbalise them. This issue is especially serious with respect to the implicit requirements, since they are obvious to the client, but unknown to the engineer. This occurs frequently, when the latter has not an in-depth knowledge about the domain. Here, the selection of the requirements elicitation techniques (see Sect. 5.3) is extremely germane, since a wise choice may allow the clients to free the requirements that they know.

The most difficult problem is however related to quadrant D, since they refer to requirements that are unknown both to the engineer and the client. An example is a law that applies in a given situation, but that both are not aware of. This situation can only be solved if a third requirements source is included in the process. This implies to exit the context that just implicates the engineer and the client to seek new sources of information, for instance, interviewing a domain specialist or consulting appropriate legislation. Anyway, at the end of the requirements engineering process, it is normal that requirements that not are not known to the two involved parties still subsist and that consequently are not reflected in the system. The aim of requirements engineering is obviously to expand the quadrant A at the expense of the other ones.

A requirements engineer must keep in mind the inevitability to constantly change the requirements. Indeed, the requirements change frequently throughout the development, because often the simple existence of a project alters the rules of the problem (Booch et al. 2007, p. 9). The contact with prototypes or even with the final version of the system forces the stakeholders to reconsider their necessities and expectations. Additionally, as the project progresses, the members of the development team gain more detailed knowledge about the domain, allowing them to better understand which are the most relevant properties for the system. Cusumano (2004, p. 154) indicates that the products developed by Microsoft have typically, at the end of the project, changes of 30 % or more when compared to the initial requirements set.

> "Progress is impossible without change, and those who cannot change their minds cannot change anything."
>
> *George Bernard Shaw (1856–1950), novelist*

Hence, the development team of a system, throughout its lifecycle, must be prepared to cope with the various changes, being able to adapt to the evolution of the stakeholders necessities. Several factors may condition this evolution, for example, the users demands, competition among companies, technology, and legislation. Thus, one must utilise mechanisms that allow documenting the system aspects more resistant to change, that thus can be used to support the other development phases. Changes in the requirements must be considered as a natural fact and not as the result of an initial request poorly formulated.

The typical approach of some development teams to try to freeze the requirements, so that based on them the development can advance, can be inadequate, although understandable since it facilitates their tasks. Alternatively, what is needed is finding the adequate mechanisms that allow changes in the requirements to be incorporated, without major perturbations in the work previously developed. Current products and systems, especially software-dominated ones, are subject to constant changes, being its functionality the most volatile aspect. Therefore, one must be concerned in reducing the volatility of the obtained artefacts, focusing on the aspects that present a stronger resistance to change. Hence, one diminishes the need to modify those artefacts and the system itself, especially when in full exploitation.

In highly complex systems, the number of requirements to be handled is normally very high, so that it becomes convenient to advance the system development in an incremental way. Thus, in each process iteration, it is necessary to choose a subset of the requirements to be tackled. Additionally, the conflicts among requirements are common, so establishing trade-offs is a task that one must promote. All these issues conduct to the need of adopting prioritisation and negotiation techniques (see Chap. 6), to find the best solutions to overcome those situations.

The requirements description should be comprehensible, both to the clients, from whom one does not expect technical competencies in engineering, and to the engineers, which generically have a reduced knowledge in the problem domain. Hence, it is necessary that the adopted notations to document the requirements are simultaneously technically rigorous and intuitively clear.

The problems and challenges associated with requirements engineering increase exponentially with the complexity of the systems to be developed. In the military, aeronautics, and medical areas, one develops many safety- and life-critical systems, such as satellites, airplanes, rockets, radars, medical and clinical devices, whose requirements documents can be organised in various volumes, each one with thousands of requirements. In systems with this nature and complexity, the produced information is so extensive that it becomes indispensable to use systematic, special and very specific requirements engineering techniques. The success of the projects of complex engineering systems depends heavily on the way requirements are handled.

All this discussion shows that the requirements engineer task is complex and multifaceted. Some authors, for instance, Brooks Jr. (1987, p. 69) and Pressman (2009, p. 119), actually consider that requirements engineering is one of the most difficult activities to be executed in the scope of software engineering. However, it is at the same time quite stimulating, given the richness of the questions that must be correctly considered and managed. It is convenient to evaluate the inherent difficulties related to requirements engineering, as well as the care to take during the tasks that are associated with it, to choose the most adequate approaches to the context under consideration. No technique used in isolation is sufficient to elicit the requirements of a system. This observation forces the requirements engineers to be able to select a set of techniques that are appropriate to the complexity and the objectives of the system at hand and to the context in which that development occurs.

4.4 Summary

Requirements engineering aims to establish communication channels between the problem holders (the users and the clients) and those that will build the solution (the engineers). The requirements engineering process includes seven main activities: (1) inception, (2) elicitation, (3) elaboration, (4) negotiation, (5) documentation, (6) validation, and (7) management. Among those activities, three of them stand out, since they are fundamental to pursuing the process: elicitation, negotiation, and documentation. In this sense, requirements engineering can be defined as the set of activities that allow the functionalities and the restrictions of the system to be elicited, negotiated, and documented, in the context of its development.

Requirements engineering is recognised by the community as having a pivotal importance for the software industry, due to the impact that its activities induce on the stabilisation and management of the whole development process. The same will happen in the industries of all other technological areas. The experience in conducting software projects made obvious that the clear definition of the requirements has a decisive influence on the quality and utility of the final system, as well as on the efficiency of the development process that was adopted. Moreover, the longer the time spent in eliciting requirements, the shorter the time that then is needed for all the development process. As a corollary, one can say that the more one invests in eliciting requirements, the cheaper is the system development. Hence, after discussing the problem, whenever possible, a series of interviews and discussions with the stakeholders, namely clients, users and domain specialists, should be organised. A project hardly constitutes a complete success, if all persons that have a stake on it do not provide their opinion (and approval) with respect to the requirements that will be incorporated.

Further Reading

There are many books that must be read by those that wish to have a complete and thorough view on all the issues related to the requirements engineering area. Various techniques valuable and accessible for requirements elicitation are presented by Alexander and Beus-Dukic (2009). The book is structured according to the intersection of nine requirements elements and five discovery contexts. Robertson and Robertson (2006) debate in detail various subjects related to requirements engineering, with an emphasis on the process. The authors use an analogy with three animals (rabbit, horse, elephant) for classifying the agility of software projects and thus discussing which approaches to use in each type of project. Lauesen (2005) addresses the requirements in general, putting together an industrial experience and a more theoretical approach. The extracts from real requirements specifications that make up about half of the book are interesting. The landmark book by Pohl (2010) is a mandatory reference, since it covers the area in an exhaustive and complete way. The book by Hull et al. (2011) is relevant, because it frames requirements engineering in the perspective of systems engineering.

The work by Jackson (2001) is relevant, because it relates the requirements subject with the problem domain, explicitly separating it from the decisions to be handled only in the solution domain. Business requirements supported by UML (unified modeling language) are the explicit target of the book written by Eriksson et al. (2004). An interesting discussion about the nature of requirements engineering can be found in (Gilb 1997). Advances in aspect-oriented requirements engineering can be found in the book edited by Moreira et al. (2013). Ballejos and Montagna (2008) propose a method to identify the stakeholders, taking into account the following dimensions: organisational, inter-organisational, and external. The method makes it possible to identify, in a significantly systematic way, all the persons, groups and organisations whose interests and needs may affect or be affected by the system under development.

Leffingwell and Widrig (2000) address the topic of requirements management. An instructive list with 25 causes for adding new requirements to a given project is presented by Weinberg (1993, pp. 403–405). The interested reader in topics like requirements interdependency and impact analysis is invited to read the following works: (Arnold and Bohner 1996; Dahlstedt and Persson 2005; Jönsson and Lindvall 2005).

Finally, the reader is directed to the main conference (International Conference on Requirements Engineering; http://requirements-engineering.org) and the top journal (Requirements Engineering, ISSN 0947-3692, published by Springer) in the field, where many scientific articles are published regularly.

Exercises

Exercise 4.1 Identify the three fundamental activities of the requirements engineering process and describe succinctly the objective of each one.

Exercise 4.2 Based on your experience, explain why there are communication problems, within the scope of the requirements engineering process, when people use technical terminology.

Exercise 4.3 For each of the seven activities (inception, elicitation, elaboration, negotiation, documentation, validation, and management) of the requirements engineering process:

1. Describe some reasons that make that activity important.
2. Indicate the challenges that must be faced to complete that activity.
3. Identify a potential consequence if the activity is not thoroughly completed.

Exercise 4.4 (Naveda and Seidman 2006, pp. 53–54) Which of the following competencies is not part of the principal role of a requirements engineer?

1. Propose a new pattern of work that improves the users performance.
2. Observe how the users work, while he interacts with them, and ask them questions about what they are doing and why.
3. Interpret the information collected from the users to better understand the essence of the work.
4. Establish, as domain specialist, the connection between the users and the development team.

Exercise 4.5 Identify some causes that make difficult obtaining a complete and coherent requirements document.

Exercise 4.6 (Naveda and Seidman 2006, pp. 19–20) An requirements team adopted the following activities to elicit the requirements of a software application: (a) creation of a questionnaire, available via web, to request information related to the proposed functionalities and to obtain suggestions of new functionalities; (b) writing of a document with all the characteristics identified by the questionnaire; (c) creation of a prototype of the graphical interface; and (d) presentation of the prototype to the top managers. Which of the following activities were completely omitted from the requirements engineering process?
(1) elicitation, (2) elaboration, (3) negotiation, (4) documentation, (5) validation.

Exercise 4.7 (Wickelgren 2012); coordinated by the instructor Distribute the participants in a circle. Write in a paper a sentence with at least 12 words, for example, "the history of the albanian town is strongly marked by the difficult survival of its people". Whisper the sentence to one of the participants, in such a way that she is the only one that hears what you say. Ask now for the sentence to circulate from participant to participant, with the restriction that each one can only say the sentence once. When the sentence arrives at the last participant, ask him to say aloud the sentence that he has received. Write that sentence on the board and share with the participants the original sentence. The group must compare the two sentences and analyse if the meaning is different. The participants should indicate if they felt they have clearly understood the sentence that was whispered to them.

This activity can be repeated, according to three variants. At the end of each one, the group should (1) compare the initial and final sentences, (2) analyse if they are similar or not, (3) try to identify the cause for (no) resemblance, and (4) suggest some ideas on how to communicate more effectively.

Variant 1: use a simpler and shorter sentence, for example, "the baron likes to ingest fried tomatoes";

Variant 2: use a sentence with at least 15 words, but now allow the participants to whisper the sentence into the ear of her neighbour more than once;

Variant 3: use another sentence with at least 15 words and, for instance, turn on a TV set in the room, to create an element that can distract the participants.

Chapter 5
Requirements Elicitation

Abstract Requirements elicitation is one of the crucial tasks of the requirements engineering process, as it allows one to discover which requirements the users want to see incorporated into the system at hand. The core content of this chapter is the description of some techniques that can be applied to elicit requirements. The chapter presents a non-exhaustive, but sufficiently representative, set of requirements elicitation techniques that can be used in engineering projects. Additionally, the chapter discusses a generic process that can be adopted for eliciting requirements and it describes some of the potential stakeholders of the system.

5.1 Process

Requirements elicitation, also designated as requirements discovery, capture, recollection, acquisition, or extraction, is a task that involves several activities that should permit requirements engineers, jointly with the stakeholders, to understand what are the requirements of a given system. In another words, requirements elicitation allows comprehending the necessities and expectations that stakeholders have with respect to that system. Many of those activities have essentially a communicational nature, which makes the origins of the associated techniques to be unrelated with the traditional engineering or science areas, but instead related with the social sciences and organisational theory.

> "Computers are good at following instructions, but not at reading your mind."
> *Donald E. Knuth (1938–), computer scientist*

This origin far away from engineering is one of the factors that makes some engineering professionals look with an unjustified contempt to these activities, due to their less technical and technologic nature. This position does not make any sense and must be discouraged and corrected, since all engineering branches and fields have important human and social dimensions. A requirements elicitation process

© Springer International Publishing Switzerland 2016

J.M. Fernandes and R.J. Machado, *Requirements in Engineering Projects*,
Lecture Notes in Management and Industrial Engineering,
DOI 10.1007/978-3-319-18597-2_5

executed in an amateur way has very negative consequences, since it seems obvious that it is not possible to construct the system desired by the stakeholders if their necessities and expectations are not known. Additionally, the requirements are frequently handled in a poor way in scenarios related to internal developments (for instance, when a given department of an organisation requests the development of a system to the engineering department), due to the erroneous perception that the proximity between the two parties eases the resolution of the doubts that may exist. Even in the case in which the client and the development team do not belong to the same organisation, the requirements are not always handled in a especially cautious way. All these situations must be strongly fought by the engineering professionals and most particularly by the requirements engineers that must insist on the need to carefully handle the requirements.

The requirements engineering process must define which is the intervention that requirements engineers must assume throughout all the development process. Actually, the activities related to requirements should not be exclusive of the initial phases of the lifecycle. Instead, the development team must adopt an approach that addresses and favours handling the requirements in all the lifecycle. It is also important to clarify the formal involvement of the stakeholders, throughout the requirements engineering process.

The definition of a universal model for the requirements engineering process is a difficult endeavour, since the interests and type of implicated persons in this process, as well as the characteristics and the framework of the system to be developed, greatly restrict the approach with respect to the tasks and activities that need to be executed. However, it is possible to define a generic process to elicit requirements that must be executed iteratively and cyclically and that can be divided, as Fig. 5.1 shows, in the following fundamental steps: (1) study the domain, (2) identify the requirements sources, (3) consult and engage the stakeholders, (4) select the techniques to be adopted, and (5) elicit the requirements from the stakeholders and other identified sources.

Despite the fact that the way one applies requirements engineering techniques is strongly dependent from case to case, they essentially cover the analysis phase, in which the requirements engineers engage primarily in two types of activities. On one hand, requirements engineers contact persons that know well the problem to identify all the restrictions that could limit the respective solution. On the another hand, they take decisions that apply to the preparation of the requirements document that describe the behaviour and the characteristics that are expected for the system. These two activities are not necessarily executed in a sequential or mutually exclusive way. While the former is characterised by uncertainty and an increase of information and

Fig. 5.1 Main steps of the requirements elicitation process

knowledge, the latter is characterised by the organisation of the ideas, the resolution of conflicting views, and the elimination of inconsistencies and ambiguities.

> "The goal of a startup is to figure out the right thing to build—the thing customers want and will pay for—as quickly as possible."
>
> *Eric Ries (1979–), entrepreneur*

5.2 Identification of the Stakeholders

Identification of the stakeholders is essential, since they are one of the most important sources of information for the requirements elicitation process, since the system is developed to satisfy their necessities and to fulfil their expectations. Leffingwell and Widrig (2000, p. 40) define a system stakeholder as someone that can be materially affected by the implementation of that system. In this book, a **stakeholder** of a system is some person that has some type of legitimate interest in that system. The term 'person' must here be understood in a broad sense, including, in addition to individuals, groups of persons and even organisations. The notion of interest is also ample and can result, for example, from utilising the system, being affected (benefited or harmed) by it, or having some kind of responsibility in relation to that system.

Despite the fact that one may think that often the number of stakeholders of a system is relatively reduced (three or four different types), a more careful analysis normally reveals a bigger number (Illustration 5.1). The identification of the various stakeholders can be performed, for instance, through the characterisation of the roles or positions that exist in the organisation. However, since some persons can accumulate, in a given period of time, several roles, the difficulty here lies precisely in distinguishing the persons from the roles they have. Imagine, for example, the director of a company that is also responsible for financial accounting. If the two roles are not separated, one risks losing one of them (in this case, most probably, the accountant), which may have negative consequences in the requirements elicitation.

There are several ways for identifying who are the stakeholders. One can, for instance, ask the client, examine the organisation chart, compare with similar products, or analyse the system context (Alexander and Beus-Dukic 2009, p. 37). One can also consult the stakeholders that are commonly found in the majority of the systems. Next, some of those stakeholders are presented and, for each one, there is a small discussion of the most relevant characteristics and roles that they have in the requirements elicitation process.

The users must be clearly identified, since it is for them that ultimately the system is going to be developed. They are one of the key beneficiaries of the system project, as they use the system for its intended purpose. The (end) **user** is any person that operates and interacts directly with the system, whenever it is in effective operation

Illustration 5.1 Some of the stakeholders for a system of lifts in a hotel

in its environment. In the case of software systems, the users are the persons that are in front of the computer screen to introduce data or observe the results. In another context, for example an automobile, both the driver and the passengers must be considered as users, despite having distinct levels of responsibility, engagement and interaction with the system user. Generically, one should give priority and preference to the requirements obtained from the users that more frequently interact with the system, than those elicited from users that only sporadically interact with it.

For each type of identified users, it is recommended to collect informations to understand if there is any particularity that must be recorded for some of the following groups of persons, while they interact with the system:

- disabled and handicapped persons;
- people with low literacy levels;
- those that do not dominate the languages used in the interfaces presented by the system;
- people with visual difficulties (users of glasses, colour-blind, partially sighted or even blind people);

- persons transporting or handling substances and objects;
- persons with reduced dexterity to interact with computer-based systems.

In a first phase, before the system is put into exploitation, the persons that are designated as users are only potential users. Only when (and if) the system is deployed and put to work are those persons effective users. In new cycles of development, for maintenance purposes, one will be able to identify users with effective experience in interacting with that system, to obtain from them relevant information to improve or correct the system.

The development of a given system implies always costs, often very high. As the economists say "there is no such thing as a free lunch", so someone must pay for it. The **client** is the entity that orders and pays for the development of a system, normally after negotiating the price with the producer or provider. This relation is often formalised through a contract that settles the obligations and rights of each party. By omission in the contract, in addition to the working system, the client should be provided with complete and detailed technical documentation, to permit the installation and (preventive, corrective, perfective, or adaptive) maintenance of the system throughout its full lifecycle. For systems that include the construction of software components, the source code of those components are also expected to be provided.

Since the system development is requested by the client, it is legitimate that she has the power to decide about or have influence on several issues, namely the scope, functionalities, and cost. Clients include, for instance, the managers, the owners of the companies or the persons responsible for a departments where the system is put to work. The clients will not always be users of the system. For example, the manager of a school may pay the development of a software application to control the canteens of the school, even if he will not interact directly with that application.

> "It is not the employer who pays the wages. Employers only handle the money. It is the customer who pays the wages."
>
> *Henry Ford (1863–1947), founder of the Ford Motor Company*

In addition to the need to ensure the continuous support and commitment of the clients during the development of a system, it is also highly advisable to engage them in the requirements elicitation activities. Actually, if the effective participation of the clients in the decisions related to the system development is not ensured, there is a large risk of the system not being accepted, when ready, because it does not incorporate the intended requirements by the clients.

A **customer** is someone who pays for acquiring a system, whenever it is available to be used. The customers are the ultimate consumers, for whom the system is rendered. It is thus recommendable to include some of them in the requirements elicitation process. In many cases, the names of the customers are already known before the development starts. This happens when the system is specifically and especially developed for a given person or organisation and, in these situations, the difference between client and customer does not exist. That reality exists in the case of the school manager that was mentioned above. In the case of a mass-market product, for example, a lift, automobile, or software application, the customers are the persons that will acquire that product when it is put into the market. It is important to understand that the act of buying does not need to be necessarily associated with a financial transaction between the customer and the seller. In the software and information systems domain, there are different business models, some of which offer the acquisition, installation, and use of the software products. In this sense, the term 'acquirer' would be more adequate, since what is really relevant in this case is the (payed or free) acquisition of products.

The characterisation of the necessities and common behaviours of the customers to which is oriented a given product defines the respective market segment. According to Osterwalder and Pigneur (2010, p. 20), a group of customers represents a distinct market segment if:

1. their necessities demand and justify a different offer;
2. they are approached through different distribution channels;
3. they need different types of relations;
4. they have significantly distinct levels of profitability;
5. they are willing to pay for different aspects of the offer.

> "The customer is not always right, but the customer always has a point."
> *Karl E. Wiegers (1953–), software engineer*

The developed products will then be sold to the customers or made available to the users. Please notice that the customers may not have the perception that they payed to utilise software. This happens, for example, whenever the software is embedded into a more ample product.

Table 5.1 shows four possible combinations for the three types of stakeholders analysed up to now. Please note that, in some cases, the same stakeholder performs several roles. In relation to the two first situations, one may say that the director of the software producer is not a customer, since he is the owner of the product, and so he does not pay to use it. It would be the same as saying that the owner of a restaurant, when he lunches there, is not a customer, since he eats for free. It is obviously a more philosophical than practical discussion, but as referred previously what is relevant here is the fact that the director has acquired the product, without necessarily implying a financial transaction. If the product eventually is not put into the market, the unique customer will be the director himself.

The three types of stakeholders mentioned before are the most evident ones, since they are present in the context inherent to the development or the exploration of most systems. More difficult is the identification of other stakeholders that are indirect users of the system or that are only affected by the business related to the system. In

Table 5.1 Scenarios of users-client-customer interaction

Roles	Stakeholder	Example
u+cu+cl	Director	• The director of a software producer decides to develop a new application for managing salaries • This application, in addition to be available in the market, is also used internally in the company that has developed it • The director is a user of that application
u cu+cl	*Chef* Director	• The director of a software producer decides to develop a new application for managing canteens • This application, in addition to be available in the market, is also used in the canteen of the company that has developed it • The *chef* of that canteen is a user of the application
u+cu cl	Shop owner Director	• The director of a software producer decides to develop a new application for managing shops • This application is available in the market • The shop owner decides to buy that application for managing his shop • The shop owner is a user of the application
u cu cl	Shop employee Shop owner Franchising director	• The director of a franchising requests a software producer to develop a new application for managing the shops of the network • The director of the software producer accepts to develop that application • The director of the franchising imposes on the shops owners the deployment and use of that application • The employees of each franchising shop are users of the application

Legend: u (user), cu (customer), cl (client)

spite of this difficulty, there are other stakeholders that should be implicated in the development process, with the aim of aiding in the identification of all the candidate requirements of a given system. Next, some of those stakeholders are discussed.

A person is considered to be an **expert** if he shows, in a given domain or subject, a deep knowledge, high-level skills and an extensive practical experience. Domain experts are useful in the requirements elicitation, since they can provide knowledge about the application domain. If, for instance, a software product in the accounting area is developed, consulting a chartered accountant or a statutory auditor is usually useful. The development of a machine tool for the textile industry can benefit from an intense interaction with textile engineers, artisans, and factory workers.

For some questions related to the development of an engineering system, in addition to the development team itself, it can be necessary to consult experts, not in the problem domain, but instead in the technological or methodological domain of that system (that is, the solution domain). These experts do not need to be involved in the technical development tasks, but can contribute with his knowledge, competencies and experience to define technical requirements, to study specific aspects of the system, or to support the decision-making of the development team.

A **developer** is a professional that executes activities that contribute to the development and maintenance of a given technical system. There are various types of developers, in function of their specialisation and the process phases in which they are more engaged. In the software and information systems domain, one can identify, for instance, analysts, software architects, programmers, or testers. In the civil construction industry, one can identify, for instance, architects, engineers, construction managers, and craft workers.

An **inspector** is someone who oversees or performs inspection over something. This category includes safety inspectors, auditors, firemen, policemen, and state agencies inspectors (for instance, entities related to the quality control area). Whenever any of these profiles is considered as relevant, it is likely that the system includes inspection-oriented characteristics and features.

A **negative stakeholder** is someone that desires that the system is not developed, neither put into operation. The attitude of a negative stakeholder can vary from a peaceful opposition to an active hostility. The identification of the negative stakeholders, although sometimes not trivial, is crucial for the development team to be vigilant to any attempt to sabotage the system development. Their presence in the requirements elicitation activities, although difficult to be achieved, may become quite relevant, namely in situations where it is necessary to identify and comprehend the personal and political relations within an organisation. In many cases, the distinct perspective of the negative stakeholders becomes useful for the discussion, which may turn their approach towards the system more neutral or even constructive.

If there are collective entities that have some kind of interaction or that may be affected by the system under development, it is recommendable to invite representatives from each one of them to participate in the requirements definition. Examples of these entities are the political parties, trade unions, professional bodies, public or governmental organisations, associations of disabled persons, religious sects, sports clubs, environmental protection groups, animal welfare associations, environmental and consumer associations, and foreign citizens or ethnic minority groups.

5.3 Techniques

The range of techniques and methods for requirements elicitation that a requirements engineer should dominate is enormous, if one analyses the broad assortment of environments, organisations, project types, and technologies that he needs to be confronted with. Therefore, the requirements engineer, in addition to all the capacities subjacent to her technical discipline, must posses the following generic competencies:

- **questioning**: to ask questions about the requirements to the right persons;
- **observing**: to witness the behaviour of the users of an existing product, system or process, to infer their necessities;
- **discussing**: to argue with the users their necessities, with the aim of formulating an understanding about the requirements;
- **negotiating**: to ease the negotiation among the users, to achieve agreed solutions about the requirements to be included, removed, or modified;

- **supposing**: to anticipate functionalities that the users may need or desire, especially when new mass-market products are bring created.

> "I wish it would dawn upon engineers that, in order to be an engineer, it is not enough to be an engineer."
>
> *José Ortega y Gasset (1883–1955), philosopher*

It is difficult to find people with such a wide and excellent profile. Partially, this reality results from the scarce education that most engineering students have in topics related to requirements.

To elicit the requirements, the analyst must know and dominate an extensive set of techniques and select and apply those that best adapt to each situation. The technique necessary, for each situation, strongly depends on various issues, like the relation between the users and the development team, characteristics of the project, or sources to be used. The difficulty lies essentially in selecting the adequate techniques and knowing how to apply them correctly (Davis 2005, p. 28).

Over the years, the requirements scientific community proposed hundreds of techniques and approaches for eliciting requirements. Hence, it is possible to identify several categories of techniques:

- **marketing**, where there is a special interest in requirements that directly contribute to the success of the system from a commercial point of view;
- **psychology and sociology**, in which one emphasises the relation between the requirements and the users, that is, the satisfaction of the necessities of the users, the individuals and social agents;
- **participative design**, where there is an active involvement of the users in the definition of the requirements for the systems that directly affect their work, following an empowerment approach quite common in many organisations;
- **human factors and human-machine interaction**, whose focus is the interaction of the users with the system, taking into consideration factors such as the graphical interface and the ease of use;
- **quality**, in which the principal interest consists in the relation between the requirements and the system quality, in such a way that the satisfaction of the users and the client is reached, by intervening in the manufacturing or development process;
- **formal methods** focused on the precision and mathematical rigour of the requirements specification.

In the specific case of the software and information systems domain, one can add two additional categories of techniques:

- **object-oriented analysis**, whose main interests is focused on the relation between the requirements and the system development process, based on the objects that exist in the real world;

- **structured analysis**, where, as happens in object-oriented analysis, the main focus is centred on the relation between the requirements and the development process, but now based on the functional processes and existing data.

> "The good analyst knows many techniques, but also when to use them and when not. He combines and modifies techniques according to specific needs."
> *Søren Lauesen (1942–), software engineer*

Alternatively, the requirements elicitation techniques can be classified according to a different reference model, that proposes three major categories, each one related to a type of source for capturing requirements (Illustration 5.2): (1) individuals, (2) groups of persons, and (3) artefacts (documents, laws, products, objects). The

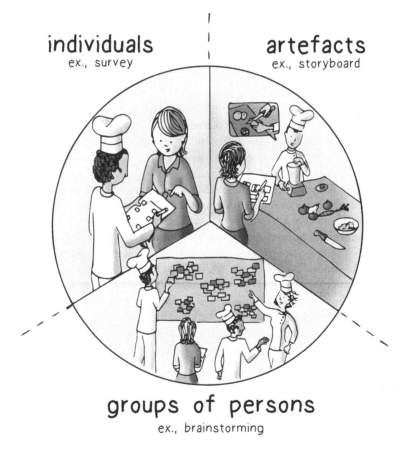

Illustration 5.2 Examples of each of the three types of requirements elicitation techniques

Table 5.2 Categorisation of the requirements elicitation techniques described in this book

Individuals	Groups of persons	Artefacts
		Domain analysis
Interview		Object-orientation
Survey	Group dynamics	Prototyping
Introspection	Cooperative work	Scenario
Ethnography		Goal modelling
		Persona

12 techniques presented in the next subsections are divided into those three categories as shown in Table 5.2. This categorisation should not be viewed as definitive and closed, but only as a simple, but effective, reference for organising the techniques. For example, it is possible to argue that observation, an ethnographic technique, can alternatively be focused on groups or that scenarios can also serve to support the discussion with the stakeholders.

5.3.1 Individuals

Interview

Due to its informal natural and ease of execution, interviews are certainly one of the most popular requirements elicitation techniques. This technique has no exact rules or formulas, since it is normally used in contexts where the human relations predominate. This fact permits great freedom to the interviewer in the conducting the interview, but may often result in low-quality results, if the interview is not focused and objective. Although any situation where one can establish a dialogue (for example, during lunch or in the corridor) may be relevant for conducting an interview, here one discusses interviews carried out in a structured and organised way, since conversations completely open seldom produce good results. Actually, an interview is not just about asking questions, but it is an activity that, as Fig. 5.2 depicts, goes through four main steps: (1) identification of the interviewees, (2) preparation, (3) conduction, and (4) conclusion.

The identification of the persons to be interviewed includes naturally some of the stakeholders (see Sect. 5.2). As a bottom line, one must interview the client and some users of the system. In the case of a product, as it is impossible to interview all the

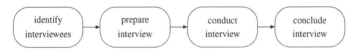

Fig. 5.2 Main steps of an interview

(potential) users, a sample that is representative of that community must be selected. The identification of the interviewees does not need to be closed before starting the interviews. It is acceptable that one adds other persons that should be interviewed, during the first interviews, by asking questions with that purpose: "who shall I also interview?" or "who else may use the system?".

> "Interviews and observation are almost always the most valuable design research technique."
>
> *Kim Goodwin (1971–), product designer*

An interview must be prepared in advance, which implies scheduling it and preparing a set of relevant questions. In special, the interviewer must provide himself with some key questions so that she can go ahead with the interviews. The answers of the interviewees can give rise to new questions that the interviewer should ask as the interview is taking place. There are various documented patterns that aim to help the interviewer in asking new questions, based on the answers that are collected, namely:

- **comparison**: when the interviewee uses a comparative adjective (for example, better, worse, faster, less useful), one can ask "compared with what?";
- **judgement**: when the interviewee uses a comparative adjective, one can ask "who says that it is better?" or "why he has that authority?";
- **generalisation**: when the interviewee says "he can (not)" or "he must (not)", one can ask "what allows (disallows) him to do that?", "why he must (not) do it?" or "what happens if he does it?";
- **universal quantifiers**: when the interviewee says "never", "always", "all" or "no one", one can ask "is it really never?" or "is it always and for all or are there exceptions?".

Conducting an interview for eliciting requirements should follow some recommendations. Firstly, the interviewer should put the interview in its context, explaining the objectives, foreseen duration, issues to be addressed and how the collected information will be processed. Whenever available, the use case diagrams (see Sect. 8.4.2) can be used as a reference for the interview. Generically, models or figures can be used to encourage the interviewee to propose modifications. Throughout the interview, the terminology of the problem that is familiar to the interviewee should be used, avoiding if possible the use of the solution domain jargon. The use of that terminology helps the interviewee to feel comfortable and keeps the dialogues in the problem domain context.

Some mechanisms that ensure a certain control over the quality of the received information should be applied. Thus, whenever the interviewer asks a question, she must hear attentively the answer and confront the interviewee with the understanding that result from hearing the interviewee answer (for example, "if I understood well, you are saying that ..."). Another possibility is to periodically obtain feedback from the interviewee about the perception that he is having about the interview (for instance, "are we progressing well?"), to make sure that it is proceeding as expected.

Another practical rule indicates that the interviewer should use less time to ask questions than the interviewees for answering them, since the main objective is to hear what they have to say. Distributions like 20–80 % are indicated as appropriate in this context, avoiding monologues from the interviewee, since that phenomenon removes control to the interviewer. Generally, whenever one hears with attention the others, one gains their sympathy and obtains useful informations that can help in solving the daily situations. The willingness to hear the others is a basic life attitude that promotes good relationships.

> "Nature gave us one tongue and two ears so we could hear twice as much as we speak."
>
> *Epictetus (55–135), philosopher*

In some cases, the presence of a third person to take notes during the interview can be relevant. Alternatively, the interview can be recorded in audio or video, if the interviewee gives explicit consent for that record to be done.

An interview can be finalised when all the questions were answered or when the time is over. Before ending the interview, one should inform about what was learnt, indicate how valuable that information is for the development team, and thank the time and effort spent by the interviewee. One may also indicate the interest in establishing new contacts with the interviewee in case there are doubts based on the analysis of the collected information or the comparison of that information with those provided by different interviewees.

As soon as the interview is finished, it is necessary to review the notes, to rewrite them in a more structured way, to reorganise the information, and to contrast it with other interviews or sources of information. The reorganised information may be sent to the interviewee, so that she can confirm its content. It is also important to evaluate how the interview was conducted to detect aspects that can be improved.

Survey

The use of surveys is common to elicit requirements, especially in the initial steps of the process. A **survey** is a technique adopted in various domains that uses a questionnaire to collect and handle the information gathered from multiple respondents.

As Fig. 5.3 illustrates, the process for applying a survey can be structured in the following fundamental steps: (1) identify the audience and the objectives, (2) conceive the questions, (3) determine the sample, (4) recruit the participants, and (5) conduct the survey.

Fig. 5.3 Main steps of the process to apply a survey

A **questionnaire** is the instrument that serves for collecting information and that is composed of a set of questions. When the same questionnaire is used for all the persons, it becomes possible to handle statistically the collected answers, something that is much more complex to accomplish in the case of the interviews, for instance. The success of the surveys is highly dependent on the way the questionnaire is conceived. Although is may seem that any literate person is capable of devising a questionnaire, reality shows that this is not always the case. On the contrary, constructing a questionnaire that is a powerful and relevant source of information is not at all easy. If the questions are not focused, if they are poorly formulated or if they appear in the wrong order, the answers that will be obtained may be not only irrelevant but even misleading. Therefore, it is important to follow some principles so that the questionnaire has the intended effectiveness.

A first recommendation consists in ensuring that the terms, concepts, and frontiers of the domain are well known both to the participants in the questionnaire and to its designers. The questionnaire must be well written, without grammatical errors and with the correct punctuation. It is equally desirable that the questionnaires are focused, to avoid information in quantity, but irrelevant, to be collected. Only a single question must be asked in each instant, avoiding several simultaneous questions. One must make sure that the answer options, when explicitly indicated, include all the possible alternatives and that they are mutually exclusive among them. The questionnaires do not permit, in many cases, the answers to be justified, neither new ideas to be explored. A solution for this limitation is including always a last alternative as an open answer. Finally, one should avoid the use of negative questions, since it is always difficult to know how to answer. For example, how to answer the question "don't you like chocolate?", if the person does not like that delicacy: yes or no?

It is also common to have questionnaires that are not totally answered and with answers poorly elaborated, in the case of the open questions. This happens since nowadays most people are always very busy with their life. The problem for unanswered questions can be tackled through the use of computer-based tools that only accept the questionnaire to be submitted, when all the mandatory answers have been introduced. However, this solution may not be desirable, since it forces the respondent to have access to a computer, which is not always possible.

The use of surveys has become quite common in recent times, since nowadays it is very easy to use web applications that permit one to rapidly distribute a questionnaire to many potential respondents. As a consequence of this easiness, persons receive very frequently requests to fill in questionnaires, which obviously diminish their willingness to collaborate.

Surveys have advantages over some other mechanisms of collecting information, being an appropriate approach in the case of persons that express themselves better in writing than orally. Additionally, surveys are relatively cheap, since they do not demand too much effort from those that ask the questions and have fixed answers that ease the compilation of the data. However, that standardisation may frustrate the users. The surveys are also limited, since the interviewees must be able to read the questions and answer them. Thus, in some cases, conducting a survey with a questionnaire may not be practical. Finally, surveys, as also happens with other explicit techniques to collect information from the persons (for instance, observation or interview), risk to get answers that do not reflect what people really think.

A joint use of surveys and interviews is possible. In that case, the survey must be used as a preliminary technique that aids in the preparation of the interviews. The doubts that the requirements engineer may have in interpreting the answers given in the survey may be addressed in the interviews. The surveys can also help in identifying aspects in which a stakeholder can more substantively contribute and that can be explored in the interview.

Introspection

Introspection is considered the most basic, obvious, and rudimentary of all requirements elicitation techniques. In its essence, it assumes that the requirements engineer defines for a given system the respective requirements based on what she thinks are the necessities and wishes of the various stakeholders. The engineer must put herself in the role of the client or the users and must reason based on premises of the type "if I were the client, I would like the product to . . ." (Laplante 2013, pp. 61–62).

> "In the martial arts, introspection begets wisdom. Always see contemplation on your actions as an opportunity to improve."
> *Masutatsu Oyama (1923–1994), founder of Kyokushin Karate*

This technique is extensively used, namely when the requirements engineers have, with respect to the clients, a deeper knowledge about the problem domain to be addressed and the processes executed by the users. The practice has shown that this technique can be used only as a starting point for the adoption of other requirements elicitation techniques. In fact, a system constructed solely based on requirements obtained from an introspection process means a lack of communication and discussion between the development team and the other stakeholders, which should not be encouraged.

Ethnography

The use of ethnographical techniques presupposes the study of the behaviour of the persons in their natural environment. The requirements engineer participates, during an extensive period of time and in an active or only passive way, in the activities that normally the users realise. He benefits from the immersion in the working environment in which the system will be used to collect informations of the accomplished operations. These techniques are quite effective whenever it is important to obtain information about how the different parts interact amongst them and how reality behaves. The usefulness of the ethnographical approaches is particularly high when there are complex social relationships among the stakeholders.

Observation is a very popular ethnographical technique and, as the name suggests, presupposes that the analyst, without directly intervening, observes and follows the execution of the current processes. It is a quite passive technique, so it is normally recommendable to complement it with the utilisation of other techniques. Additionally, it is a technique that turns out to be expensive and that forces the analyst to a large effort and a high abstraction capacity to interpret and generalise the activities that he observes being performed.

Anonymous observation in public places raises the need for informed consent, an ethical consideration that does not exist in interviews (Goodwin 2009, p. 186). If the users realise that they are being observed, there is a high risk that they execute the activities in a (more careful and controlled) way when compared with the one that would occur in a normal situation. Under these circumstances, the observation efforts consider phenomena that may be deviated from their normal reality, distorting the results and diminishing thus the effectiveness of the technique. In some cases, it may make sense to video record the activities for further analysis.

> "In the fields of observation chance favours only the prepared mind."
> *Louis Pasteur (1822–1895), microbiologist*

Another ethnographical technique is *apprenticeship*, in which the analyst learns and literally executes the activities under the supervision and control of an experienced user (Beyer and Holtzblatt 1995). This technique is especially relevant when the analyst has no knowledge about the problem domain or when the users feel difficulties in explaining or verbalising the activities that they perform.

5.3.2 Groups of Persons

Group Dynamics

The use of techniques based on group dynamics is quite common and recommendable in requirements elicitation contexts. If a relatively high level of participation and a certain dynamics of cooperation among the participants in the sessions are reached, these techniques can constitute a very effective form of capturing the requirements. However, this type of sessions is normally hard to organise, especially in projects with a high number of persons. If it is not possible to ensure the regular presence of all the stakeholders representatives, the effectiveness of this technique may not be as high as expected. Additionally, managing the sessions is also not an easy task in most cases, since it requires experience and competencies hard to find. In particular, the analyst, as moderator of the sessions, must avoid that some participants try to dominate the discussions and make sure that all persons feel comfortable to express their ideas and opinions in a sincere and frontal way, with no fear of being a target of possible retaliations in the future.

Brainstorming is a group dynamics technique that eases the generation of ideas. A brainstorming session congregates a group of 5 to 12 persons that suggests and

Fig. 5.4 Main steps of a brainstorming session

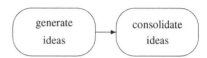

explores as many ideas as possible for the system to be developed, without criticising or judging those ideas. As Fig. 5.4 depicts, each session includes two steps: idea generation and idea consolidation.

In the first step, the participants are encouraged to generate ideas, without discussing the value of each one. Hence, ideas that may seem foolish or nonsense are not rejected (at least in this step), since the generation of a high number of alternatives of formulating the problem increases the chances to find the "right" idea. The basic principle of this step assumes that the group effect and the absence of criticisms are sufficiently inspiring and motivating to free the group members of any mental blocks, allowing them to create new ideas, even if those ideas seem weird, strange, or crazy. Sometimes, the group generates quite interesting ideas, based on ideas previously presented. This step requires the generated ideas to be visible and available to everyone.

> "If you want to have good ideas you must have many ideas. Most of them will be wrong, and what you have to learn is which ones to throw away."
> *Linus C. Pauling (1901–1994), chemist*

In the consolidation step, the ideas are discussed, reviewed, and evaluated. The objective is to organise the ideas with the aim of increasing their chances of utilisation. It may happen, for example, that some ideas are significantly identical, being necessary, in that circumstance, to combine and rewrite those ideas to capture their common essence. Many of the ideas can be certainly useless or considered too weird, but they however fulfil their purpose if they inspire the creation of ideas with higher interest/utility. The remaining ideas, considered as interesting, must be discussed in detail with the objective of classifying them according to an order of preference. At the end of the session, the moderator documents the ideas considered relevant and identifies their order of priority. Those ideas may require further elaboration with more detail and consequently one needs to apply other requirements elicitation techniques, like the ones described in this section.

According to Rossiter and Lilien (1994), when the brainstorming technique is used, six principles must be followed in order to obtain high quality and more creative results:

1. Clear instructions are necessary to maximise the creative ideas: (a) there is no place for criticism, (b) crazy ideas are valuable, (c) the quantity of ideas is important, (d) the ideas must be combined and developed.
2. The problem must be specific and sufficiently difficult.

3. Persons working alone, and not in groups, should generate the initial ideas. The ideas generated must be registered without revealing its author and next read and commented by everybody.
4. The interaction within the group, composed of those that previously have generated the initial ideas, must be promoted, to allow those ideas to be refined, combined, and enhanced.
5. The best idea must be selected by individual votes.
6. The decision must take a relatively short period of time (typically, less than two hours), but it should also not be taken in a rush.

It is recommendable that all the development projects have at least a brainstorming session, at the beginning of the development process, before taking the major decisions. This technique has a relatively low cost and it is easy to execute. Additionally, it stimulates imaginative and creative thinking and can avoid the tendency to limit very early the shape of the system.

Another example of a technique of this type, commonly used for market research, is *focus group*. The preparation and conduction of the discussion sessions in group are, in many aspects, similar to those that one can observe in interviews (for instance, preparing questions in advance and provide feedback of what one hears). The focus group technique consists in gathering, during 60–90 min, a group of persons to collectively provide their opinion about a relevant topic related to the development or maintenance of a system. The group discusses, in a relatively planned way, the topic of interest (Langford and McDonagh 2003, p. 2). The moderator can use auxiliary materials, like prototypes or movies, to stimulate the opinion of the participants. The discussion is observed and summarised. If several focus groups are organised, each one with distinct participants, the various summaries allow one to come up with useful conclusions about the requirements of the system under development. This technique has, when compared to interviews and surveys, the advantage of easing the discussion and the participation, since the answers of a participant can be complemented by another one, enriching thus the information. It proves appropriate for developing a new system, especially when the development team knows little about it.

Another technique in this category is JAD (joint application design) that aims to ease the collaboration within a group composed of users, problem domain experts and engineers. The objective is to organise JAD sessions in which the participants jointly discuss both the problems that must be solved and the respective available solutions. The idea behind this technique is based on the assumption that it is possible to rapidly take correct decisions with respect to the requirements of the system.

"Alone we can do so little; together we can do so much."

Helen Keller (1880–1968), political activist

There are some similarities between brainstorming and JAD, although in this technique the principal objectives for the system are assumed to be already established before the stakeholders meet. It is also common that the JAD sessions are more structured than the brainstorming sessions, since the former are usually defined by a series of steps and actions to be followed and the roles of the various participants are well characterised.

The success of the techniques of this type depends mainly on the dedication and effort of the participants, so it is crucial how they are selected to participate. Generically, the identification of the stakeholders (see Sect. 5.2) can help on the choice of the participants. In particular, to obtain valuable results, it is often fundamental to ensure the participation of persons with some creativity and imagination, but simultaneously with deep knowledge and expertise in the relevant domains.

These techniques also require that someone takes over the role of moderator or facilitator, to make sure that the group is objective and all the programmed issues are effectively addressed during the sessions. Even if, in the common sense, moderator and facilitator are concepts generally seen as synonyms, in a more rigorous way, there are some differences between the two roles. The *moderator* should be seen as a manager of the produced information over the sessions by the implicated groups. He/she may also manage the time and produce some alerts with respect to the principal objective of the discussion under consideration. The *facilitator* corresponds to a more active version of the moderator, taking responsibility for the success of the group work, without assuming position with respect to the subjects being discussed. The facilitator stimulates the most passive participants, restrain the most impetuous participants, and manages possible conflicts within the group.

Cooperative Work

In contexts where the collaboration among the members of a group cannot happen in the same space and at the same time, the group dynamics technique mentioned before are not viable. A situation, in which a group resorts to the interaction among the participants in an asynchronous and distributed way (that is, that can occur in different moments and places), is referred to as *cooperative work*. This possibility is very powerful, since in many situations it is hard to guarantee that the persons are able to meet in the same place at the same time.

Cooperative software, also designated as *groupware* or by the acronym CSCW (computer-supported cooperative work), is a software application that eases work in group executed cooperatively. A tool of this type permits a group of persons to work in a collaborative way, for example, editing the contents of a given document.

The use of cooperative software can be framed in all types of projects, but seems to be specially adequate in systems that include a high number of persons. The success of this technique depends on the participation of the persons, but it has been evident that many people are willing to dedicate some of their time in electronic forums to provide their opinion about a given subject or to contribute to the collective construction of some enterprise. The enormous success of the Wikipedia suffices to show this fact.

"Great discoveries and improvements invariably involve the cooperation of many minds."

Alexander Graham Bell (1847–1922), telephone inventor

The use of this type of software applications can be justified by several reasons, as indicated by (Davis 2005, pp. 34–35): (1) permit anonymous contributions, (2) permit the participation of persons geographically dispersed, (3) allows written, and not oral, contributions, (4) facilitate information to be recorded in an electronic format, (5) shield shy persons.

Examples of cooperative software are the blogs and wikis. In spite of the fact that nowadays wikis are used in many distinct contexts, it should not be a surprise their use to support requirements engineering activities. Firstly, because the origin of wikis is precisely as tools to facilitate the communication among the persons implicated in a given software development project. Secondly, because being analysis the phase in which it is necessary the greatest level of communication, the use of tools that support that need is useful. A wiki permits all the participants to add new informations or to edit and change informations previously introduced. Its utilisation can be quite pertinent, for instance, in contexts of global software development. This software engineering trend is motivated by the desire (1) to distribute the development teams, that is, to put their members in different places of the planet, in order to profit from the 24 hours of the day, (2) to maximise the use of the resources, and (3) to increase the proximity to the final client.

5.3.3 Artefacts

Domain Analysis

A way of capturing requirements for a given system consists in analysing documentation and studying existing systems. The study of these systems presupposes, in some cases, analysing those that are located upstream and downstream in the business process chain, as well as examining for competing systems or with similar purposes. This technique is important to obtain a larger knowledge about the problem domain (*cf.* step 1 in Fig. 5.1). The objective of domain analysis is not examining a specific system, but instead the domain in which it is located, to identify the common elements for solving the problems that are applicable to all systems in that domain. In some cases, it is expectable that the producer of the system has a domain model (see Sect. 8.4.1). In that sense, domain analysis can be considered an activity in continuous execution, whose objective is to provide analysis patterns that can be used for developing any system in that domain. Domain analysis is part of a wider framework that is designated as *domain engineering*. The objective of domain engineering is to gather, organise and store artefacts related to the construction of systems in a given domain, as well as providing appropriate mechanisms for reusing those artefacts when one is building new systems (Czarnecki and Eisenecker 2000, p. 20).

In the software and information systems domain, the most recent tendencies show that there is a reduction in the number of developments from scratch, in contrast to the growing in the number of systems in operation that need to be modernised and altered. To prove this fact, one must remember the adaptations that were introduced in many software systems, due, for example, to the alleged year 2000 bug and the transition for the new European currency (Euro). Thus, frequently organisations feel the need to correct or improve a working system. Although it is admissible to frame this activity in the maintenance phase, it seems more appropriate to consider that a new development iteration will be executed, if new technologies or approaches are needed. In this situation, the behaviour of system, as well as its documentation, if available, can and must be used to base the analysis to be conducted for the new development. This technique is known as reverse engineering.

Consulting and analysing documents is a technique based on searching requirements in documents (reports, diagrams, models, manuals, minutes of meetings, surveys, interviews records). Obviously, it is a technique that can only be applied if there is already a legacy system. The utilisation of this technique makes sense if the systems need to be changed or improved. This technique has obvious limitations, so it is recommendable to complement it with other requirements elicitation techniques. In a given organisation, the use of this technique will be possible in the future, if the artefacts for the current projects are stored.

Object-Orientation

Object-oriented analysis, extensively used in the software and information systems domain, studies the requirements, in the perspective of the real-world objects (and of their classes) considered in the vocabulary of the problem domain. The user requirements should be expressed in the terminology of the problem, defining clearly which functionalities the system is expected to make available. In other words, user requirements must be defined according to an operational perspective. As the survey made by Wieringa (1998) suggests, over the years many object-oriented analysis methods were proposed, which shows the enthusiasm established around this approach. Their advocated refer frequently some of its advantages, especially that it represents a more natural way of studying systems, since it provides a direct correspondence between the real-world objects and the elements of the model that result from the analysis. Object-oriented analysis encourages the analyst to focus on what the system must do, instead of how it should be built. It permits also a softer transition for the design phase and promotes the reuse of models created in the context other systems.

Finding (or identifying) the real-world objects related to a system is not a trivial activity, despite what one could think at first sight. It is an issue that produces in most cases a lot of discussions and that can introduce many divergences within the development team. The objects can be identified in many forms by the engineer familiar with the problem domain, and choosing an adequate representation is a development task can be hardly realised without humans. A possible strategy used for identifying the objects associated with a given system, as well as their attributes

Fig. 5.5 Main steps of the process, suggested by the OOA method, to analyse systems according to an object-oriented approach

and operations, consists in extracting the nouns and the verbs from the problem description in natural language. The objects and the attributes correspond to the nouns and the operations to the verbs (Abbott 1983). This strategy must be used with caution, since it is always possible to transform a noun into a verb and vice-versa (for example, "the game is interrupted at 3pm" can be transformed into "the interruption of the game happens at 3pm").

> "Object-oriented analysis is based upon concepts that we first learned in kindergarten: objects and attributes, wholes and parts, classes and members"
> *Edward N. Yourdon (1944–), software engineer*

For example, the OOA method, proposed by Coad and Yourdon (1991, p. 34), is exclusively focused on object-oriented analysis and is divided in five steps, as illustrated in Fig. 5.5: (1) localisation of classes and objects, (2) identification of structures, (3) identification of subjects, (4) definition of attributes, and (5) definition of operations (services, in the terminology of the method). These steps must be seen as activities, which means that their order is not fixed, although the one shown corresponds to the most common application of the method.

In the first step, the aim is to identify structures, other systems, devices, events, roles, operational procedures, organisational places and units that can be considered as possible objects or classes. In the second step, one identifies the (generalisation/specialisation or composition/aggregation) relations between the classes and the objects. The third step permits us to divide the problem domain in subproblems. For that purpose, the classes and the objects with the highest levels are examined to transformed them in candidate subjects. In these two last steps, the attributes and operations of the classes are defined.

Prototyping

Sometimes, the client just defines some generic objectives for the system, not indicating with detail its functioning. In a situation like this, a prototype-based approach can be the most adequate choice to support the requirements elicitation process. This approach assumes an iterative process, whose structure is represented in Fig. 5.6. This model can be considered a reinterpretation of what is presented in Fig. 2.4.

The prototyping process tends to be iterative, repeating cyclically the requirements elicitation, the construction of the prototype and its validation from the users. Each

Fig. 5.6 Main steps of the process suggested by systems prototyping

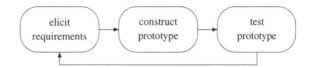

iteration allows stakeholders to gain a more solid understanding of the requirements, namely those that were established in previous iterations.

Prototyping starts with the requirements elicitation; in this task, the client and the analyst meet and define the known requirements and identify areas in which further intervention is needed, through the refinement of the requirements previously considered and definition of new ones. After this step, one tries to construct rapidly and with no special attention to the details, the **prototype**, defined as a easily modifiable and extensible working model of the system, which provides users with a representation of its key parts before implementation (Connell and Shafer 1989, p. 23). A prototype is usually a rudimentary and temporary version of the system, since only part of its aspects are considered, namely those that are related to the external view.

The built prototype is evaluated by the users and is used for detailing and complement the requirements of the system. The evolutive nature of prototyping is due to the fact that prototypes are gradually modified to better satisfy the expectations of the stakeholders, with the objective of allowing the development team to better comprehend what the system must provide to its users.

The prototype serves, among other things, as a mechanism for capturing the requirements. As soon as one considers that the requirements of the client are clearly understood, usually the prototype is abandoned.[1] Then, the development process is started, taking into account the requirements identified with the help of the prototype and following a process in accordance with a given development process model (see Sect. 2.4).

> "It's a prototype–not the Mona Lisa."
>
> *Todd Zaki Warfel (1971–), product designer*

Different forms of prototyping, with quite distinct levels of sophistication, exist, but all of them must actively engage the users in the definition and validation of the requirements. When new systems are developed, the use of prototypes is highly recommendable, since it permits the users to have contact with a version of the system, much before it is completely finished and in effective operation.

One of the main advantages of prototyping lies in the possibility of initiating the development, even when all the intended requirements for the system are not totally captured. Additionally, prototyping permits the construction of something that works,

---

[1] There are other perspectives about prototyping, but only this one is assumed here for simplicity in the discussion.

providing the users with an idea of how the system will behave. That would not be possible so easily if the users were confronted with a paper-based specification.

In particular, prototyping is very useful, in the software and information systems domain, for analysing issues related to the graphical interface and the usability of the system. Static prototypes, based for instance on paper/cartoon, allow the analyst to act as the computer and, in front of the users, to move the interface elements in response to the actions of those users. The difficulties, doubts, questions, and comments of the users with respect to what is happening are recorded. Executable prototypes, i.e., built in software technology, allow a higher level of realism, but it is necessary to not create versions too sophisticated, since a prototype must possess a level of complexity and cost significantly lower than the final system.

As disadvantages, this technique presents the fact that both the client and the development team tend often to forget that the prototype is not the final version of the system, but instead a medium to ease requirements to be elicited. The prototype is developed, ignoring a set of factors, as a way to simplify its construction and with the principal purpose of capturing the requirements. However, after interacting with the prototype, the stakeholders can be tempted to think that with some marginal alterations to the prototype can be enough to obtain a version of the system with reasonable performance levels, which hardly ever correspond to the truth.

Scenario

Scenarios constitute a very popular approach for capturing requirements and are considered a mechanism that promotes the communication between the development team and the stakeholders. Succinctly, a **scenario** is a (small) story that describes the functional behaviour of a system and illustrates a specific sequence of actions and events necessary for its execution. A scenario represents necessarily a partial and reduced story that describes the behaviour of a functionality of a system and portrays a specific sequence of actions and events necessary for the execution of a set of functionalities. So, various scenarios are necessary for describing the different alternatives of execution of those functionalities. The scenarios are a technique that aids the stakeholders to go beyond what is obvious, helping them to discover the requirements that they really want and even to suggest creative and innovative ideas (Robertson 2004).

Within the requirements engineering, the term 'scenario' is essentially used with three different meanings (Sutcliffe 2002, p. 121):

1. A story or an example of events, taken from the experience with the real world.
2. A single path through a given behavioural model, typically a use case.
3. A future view of the system to be developed with sequences of behaviour.

Considering the three alternatives mentioned before, the scenarios are expressed in different formats. For the first alternative, one uses narratives in free text, that can be complemented with images to help in the creation of the story context. The second alternative uses either sequence diagrams (see Sect. 8.4.4) or ordered lists with the textual descriptions of the steps needed to be executed. This is an approach that is followed by the object-orientation community that defines scenario as a specific sequence of actions that illustrate the behaviour (Booch et al. 1999, p. 466). In the last option, it is normal to resort to *storyboards* or even to animations or simulations. In this case, the scenario presents similar characteristics to those of a prototype.

Storyboards, largely used by the film industry, are similar to comics. They are graphical organisers, where illustrations and images are arranged in sequence with the purpose of pre-visualising a film or animation. A storyboard aims to mark the most important passages of the story of a given film, so that the persons interested in the process of filming do understand the sequence and the rhythm of the scenes. In the software domain, storyboards are sequences of images used for showing the fundamental steps in the interaction between the system and the users. Those sequences allow capturing the causality relation between the actions of the users and the system reactions. Hence, stakeholders, by visualising those interactions, are able to imagine how the system works, easing their capacity to discuss the respective requirements. Similar results can be obtained with the utilisation of sequence diagrams, although this notation has a stricter format, that may limit the creativity used to represent the scenarios.

> "I only storyboard scenes that require special effects, where it is necessary to communicate through pictures."
>
> *John Boorman (1933–), cineast*

An approach that is usually followed consists in firstly describing the normal or typical scenario, that is, the specific situation in which everything runs as expected. Afterwards, the variants to the typical scenario begin to be analysed and described, as well as the exceptions that may occur. The number of variants can be very high, so if there is no control, one can become paralysed in this process, collecting indefinitely scenarios. The requirements appear in a natural way, since the users are asked to describe simple scenarios of the use of the system. As more scenarios are described, more steps of the functionalities are discussed, enriching thus gradually the comprehension about what the system must offer.

Goal Modelling

Requirements elicitation techniques are naturally focused on capturing the requirements from the stakeholders. Goal modelling must be complemented by other techniques, since it has a distinct focus. This technique addresses pre-elicitation steps, aiming to help the stakeholders to understand which alternatives there are for using the systems and what are the implications that these alternatives have for the different stakeholders.

For such, it is necessary to identify the objectives of the organisation that will be satisfied by the system being developed and the reason why it is being commissioned. The focus is on comprehending the "whys" that are behind the construction of the system and not so much in describing "what" it can do. Therefore, it is necessary, among other things, to analyse in detail the organisation where the system will be deployed. Here, it is relevant to understand the domain in which the organisation operates, its mission, vision, and values.

A **goal** is a high-level purpose (or intention), often related to the business or the organisation, that a stakeholder wishes to see reflected and fulfilled in the system at hand. A goal is a form of capturing the reasons by which it is necessary to develop the system. The objectives are expressed by sentences in a natural language that indicate the purposes to be achieved by the system (Dardenne et al. 1993).

> "When it is obvious that the goals cannot be reached, don't adjust the goals, adjust the action steps."
>
> *Confucius (551 B.C.–479 B.C.), philosopher*

Goal modelling is a technique that structure the objectives hierarchically. The objectives of the highest level are related to the ones in the lower levels (also designated sub-objectives). These relations may use relational operators (for example, 'and' and 'or') and must be properly justified. This process is repeated in as many levels as those necessary to reach the individual requirements of the system (at the lowest level).

A similar technique to this one is *task analysis* that serves to understand the persons and how they perform their work. A task is an objective complemented by a ordered set of actions (Benyon 2010, p. 252). Task analysis resorts to an hierarchical top-down approach, in which high-level tasks are decomposed in subtasks and these in detailed sequences, until all the events and actions are identified and described (Richardson et al. 1998). With this approach, one obtains a hierarchical list of the tasks performed by the system and its users. It is a technique extensively used for defining the human-computer interface of existing systems (Rogers et al. 2011, p. 383).

The combined use of goal modelling and scenarios is suggested by some researchers, for instance, Rolland et al. (1998) and Kim et al. (2006). These approaches are particularly useful when only the high-level objectives of the system are clearly known and there is no perception neither knowledge with respect to the specific details of the problem to be addressed and its possible solutions.

One of the disadvantages that can result from using this technique is that the errors made in the definition of the high-level objectives have generally an enormous impact both on the sub-objectives and the requirements that originate from them. Thus, any change in the objectives has a very high cost, which unfortunately creates some reluctance to carry out those changes.

Persona

The persona technique, created by Cooper (1999), is common in the advertising area and, in the last times, has gained some popularity in the requirements engineering area, especially in the context of product or service development. A **persona** is a fictitious person that represents an important type of the users of the product under development. A persona is an archetype of the persons that are part of the target audience, i.e., those persons that may buy or use the product. A persona should thus be conceived to represent those persons, in what is essential and distinctive. The personas are a technique of market segmentation. The idea behind the personas is that it is better to understand and completely satisfy the necessities of the critical few than to poorly meet the needs of many (Lidwell et al. 2010, pp. 182–183). The use of personas must occur in the initial phases of the creation of a new product, since it offers hints and pointers that orientate the engineers about the necessities and concerns of real users.

To create the profile of a persona, at a minimum, the following elements must be defined:

- The **type** of the persona.
- A **photo**, so that the persona has a face.
- A **name** to permit one to call the persona by his name. Sometimes, some strong characteristic of the persona is added, like for example "Leonard Turner, the innovative lawyer".
- Some **personal data**, like for instance the academic qualification or the favourite hobbies, to frame the persona in his daily live.
- A **narrative** that tells a story of the persona. A list of items is usually not enough to transmit so well what is intended.
- The **final objectives** that the persona aims to achieve and that the product is expected to satisfy.

There are no hard rules for using this technique, but it is common to include some of the following data: sex, age, civil status, family, local of residence, level of education, professional profile, salary, life style, values and principles, motivation, expectations, necessities, context for using the product. The personal data are relevant for facilitating the memorisation of the persona, but one should not exaggerate, since if those data are too numerous and detailed they may deviate the attention from the essential behavioural data that the persona represents. A famous person, like Barack Obama or Cristiano Ronaldo, should not be used (not even a known name), since that normally works as an element that restricts the persona to that real person. When the product under development fits in the software and information systems domain, it is important to indicate, for each persona, his competencies with computers, namely if it is easy (or not) for him to install, use, or configure software applications. Table 5.3 shows an example of a persona with various data that concretise the elements just mentioned.

> "Persona is a mask that is adopted by a person in response to the demands of social convention."
>
> *Carl G. Jung (1875–1961), psychiatrist*

This technique aids the development teams to focus on the users and their necessities. Hence, the team can concentrate on satisfying the requirements of that person,

Table 5.3 Example of a persona

Susan Taylor, the always-connected bank clerk

Age: 28 years old
Civil state: single
Academic qualifications: Graduated in business management from
University of Surrey
Profession: bank clerk
Salary: EUR 1.500/month
Residence: 1-room house, located in Slough (32 km west of central
London)

Life style: Susan likes to go out in the evening with her friends, especially during the weekends. She loves going to the cinema and shopping. In particular, she cannot resist buying new shoes, having more than one hundred pairs in the wardrobe. Some pairs were only used once or twice. She would like to move to a more chic area with more educated neighbours than those that now she puts up with. She is professionally punctual, but rarely arrives on time in her personal meetings, since she takes a lot time to ready. She is looking for a boyfriend for a serious relationship, since she wishes to be a mother before 32 years old.

Context of using the product: Susan cannot live without her smartphone and she is constantly reading and writing email messages and consulting the pages of her friends in the social networks. She uses applications for the smartphone that allows her to be aware of the most recent songs in market. She likes to hear the current *hits* and knowing which concerts will be organised in the coming weeks. She already attended with her girl friends some summer festivals and she would like to repeat the experience.

Objetives:

1. be informed about concerts that include artists that she appreciates;
2. be able to forward those informations to her friends through the social networks;
3. to receive suggestions about recent songs that can please her, based on her tastes.

Image courtesy of stockimages at www.FreeDigitalPhotos.net

instead of trying to find all the possible requirements of all the persons that can have interest in the product. It is a technique especially useful when the number of users is very high or when they are not available, which happens, for example, with software products to build from scratch.

It may be recommendable to create more than one persona for the same product, namely when it is directed to a target audience composed by different profiles, like, for instance, persons in distinct age-groups, with differentiated levels of education, and with diverging degrees of capability in using the technologies. However, some authors suggest that the number of personas should not be greater than three, since a bigger number is a strong indication that the target audience of the product may not be sufficiently well identified. Other authors, like for instance Goodwin (2009, p. 238), assert that there is no magical number and that the number of personas can go from 2 up to 25, or even more.

According to Adlin and Pruitt (2010, p. 2), the process associated with the use of personas goes through five steps, as depicted in Fig. 5.7. Succinctly, the process starts with a clear identification of the problem to be addressed and the informations available for its characterisation. Next, those informations are systematically

Fig. 5.7 Main steps of the process to use personas

processed to create the relevant personas for the problem at hand. It is then necessary to plan how the personas will be introduced within the development team and the other departments of the software producer. Next, the personas are used for aiding in performing the various tasks associated with the development and commercialisation of the product. Finally, the success of the personas is evaluated and the respective use in future projects is discussed.

5.4 Summary

Requirements elicitation is a task that involves various activities that allow to identify which are the requirements for a given system. The requirements engineering techniques cover essentially the analysis phase, in which the requirements engineers dedicate their efforts to contact the persons that know well the problem, identify all the restrictions that can limit the solution, and decide how to organise the requirements document, that describes the system behaviour.

The identification of the stakeholders is an important task, since they constitute one of the most important sources of information when eliciting requirements. A system stakeholder is a person that has some sort of legitimate interest in that system. The notion of interest is wide and can result, for example, from the fact of using the system, having some benefit or prejudice due to its existence, or being somehow responsible for it. The stakeholders discussed in this chapter include the users, clients, customers, technical experts, developers, inspectors, and negative stakeholders.

The analyst is expected to know and dominate a large set of techniques and to be able to select and apply those that best adapt to each project. This chapter describes 12 requirements elicitation techniques, that are divided in three major categories, each one related to a type of source for capturing requirements: (1) individual persons, (2) group of persons, and (3) artefacts.

Further Reading

To find out more about requirements elicitation techniques, we recommended reading (Zowghi and Coulin 2005), which presents an excellent summary and comparison. Goodwin (2009) discusses, in a product design perspective, various techniques that can be useful for requirements elicitation. Macaulay (1996) presents some elicitation techniques divided in three groups: (1) approaches for capturing political and organisational requirements, (2) approaches based on group sessions for facilitating the communication among the involved persons, and (3) interactive approaches for understanding the work of the users.

Some ideas about how to conduct interviews can be obtained in journalism and social communication books, for example, (Fleisher and Gordon 2010). The example of an interview, presented by Goodwin (2009, Chap. 8, pp. 155–181), deserves to be analysed. To learn more about how to prepare surveys and questionnaires, many books can be consulted, for instance, (Brace 2013; Gillham 2008).

There are a lot of books about how to develop software with objects. Mandatory books are (Booch et al. 2007; Budd 1997; Coad and Yourdon 1991; Meyer 1988; Rumbaugh et al. 1991). These object-oriented development approaches are, from an historical point of view, a natural evolution of the structured analysis approaches, since they adopt many of their principles and introduce others with the aim of solving some of their problems and weaknesses. Although the use of structured approaches has no longer the same popularity that was witnessed in the 1980 decade, the interested reader should consult some of the landmark works of this methodological trend that left some interesting ideas, concepts and practices on the way software is developed nowadays. Thus, (DeMarco 1978; Gane and Sarson 1979; Page-Jones 1980; Davis 1983) are worth reading.

Different types and forms of prototypes, including active and passive, high- and low-fidelity, horizontal and vertical, are discussed by Constantine and Lockwood (1999, Chap. 10) and (Stevens et al. 1998, Chap. 10). Another interesting book, authored by by Warfel (2009, Chap. 10), provides a mixture of theory with many practical guidelines on prototyping for user experience.

There are some goal modelling approaches. The most popular are KAOS (Dardenne et al. 1993) and i* (Yu 1997). A systematisation of the available approaches, made by Horkoff and Yu (2011), provides some clues about which criteria to use for selecting the most appropriate approach in a given situation.

Gottesdiener (2002) presents and discusses how *workshops* can be used as a technique for eliciting requirements based on group dynamics. A workshop is a structured meeting in which a group of stakeholders carefully selected works jointly to define, create, and refine the user requirements. The book is useful not only for software-intensive systems, but also for other types of engineering systems.

The book organised by Alexander and Maiden (2004) is a reference in the subject of scenarios and includes various techniques that can be used in business contexts. The utilisation of storyboards in cinema is explained by Rabiger and Hurbis-Cherrier (2013, pp. 307–308) and Simon (2007).

Exercises

Exercise 5.1 (Naveda and Seidman 2006, pp. 63–64) Imagine that a software engineer, who concluded recently his degree at a given university, is leading a requirements engineering team for a project to improve the software application that permits students to enroll and register in degrees offered by that university. Which of the following requirements elicitation techniques are adequate for capturing the typical and atypical activities involved in the use of the application?
(a) Observation, (b) Prototypes, (c) Interviews, (d) Surveys.

Exercise 5.2 (Naveda and Seidman 2006, pp. 69–70) For the system indicated in the previous question, during the requirements elicitation process, some students were interviewed. They have indicated the functionalities that they would like to see incorporated in the final solution. Afterwards, the client has requested to remove some of the requirements proposed by the students. Which of the following arguments is the **less** strong for justify the removal of those requirements?

1. The requirements from the students are not representative of those from the student population.
2. The requirements from the students are ambiguous and cannot be tested.
3. The requirements from the students are contrary to the interests of the client.
4. The client does not consider the students as system stakeholders.

Exercise 5.3 (Naveda and Seidman 2006, pp. 51–52) Which of the following arguments is the **strongest** to justify the use of the observation technique in a company?

1. Direct interaction with users permits a continuous discussion about the various forms of work.
2. Observation permits one to see not just the normal workflow, but also less typical situations.
3. Observation is a traditional technique for capturing requirements and the company has experience in using it.
4. Observation aids in the observer/observed interaction, when they exchange ideas in real-time.

Exercise 5.4 Suppose that the analysts of a software product project have a reduced knowledge about the respective domain. Which requirements elicitation techniques are the most appropriate in that case?

Exercise 5.5 Explain the main reasons why the combined use of ethnographical techniques with prototyping is useful for eliciting requirements.

Exercise 5.6 Identify the problems that are present in the following questions that are part of a questionnaire for collecting information about a software application:

1. Why do you prefer the menus on the left rather than the right side?
2. Do you normally use the same password on different systems?

3. Where do you download email messages?
   ☐ at home   ☐ at the office   ☐ at school
4. When you go to the canteen, do you drink orange juice and eat soup?
   ☐ yes   ☐ no
5. How many hours did you sleep last night?
   ☐ 9–12   ☐ 6–8   ☐ <6   ☐ >12
6. How many email messages do you receive on average each day?
   ☐ <30   ☐ 30–50   ☐ 50–70   ☐ >70

Exercise 5.7 Identify all the stakeholders for the following systems:

1. lifts in a hotel (see Illustration 5.1);
2. commercial plane;
3. train station;
4. web application to reserve rooms in hotels and B&Bs;
5. web application for buying tickets for musical concerts and cultural events.

Exercise 5.8 Select an engineering system in which you are currently working. If it is not possible to select a system, imagine that you are involved in a project for developing a system for a logistics company (in which trucks are used). The objective of that system is to permit truck drivers to receive radio messages with delivery instructions.

1. List the types of stakeholders in the system.
2. List the job titles and roles of the persons that you consider relevant to interview for each type of stakeholder identified in question 1.
3. If you have considered a real project, speak to the persons listed in question 2; ask them whom they interact with (for instance, clients and suppliers) to achieve their business objectives. Exclude persons that not do not seem relevant.
4. Repeat the previous steps, until the list of stakeholders stabilises.

Exercise 5.9 Consider the following situations in which you were supposed to apply the observation technique. Answer and justify the questions.

1. You are developing a product to help a pastry business. Do you think that it is ethical to go to a competitor, sit down at a table, ask for a coffee, and observe the customers and the employees?
2. You are developing a game for teenagers. Do you think that it is ethical to go to a school to observe how they behave? And what is your opinion about approaching some of them and asking some questions?
3. You are developing a product for a hospital. Do you think that it is ethical to observe the behaviour of the persons in the waiting room? do you think that it is ethical to attend the appointments that patients have with their doctors? And what is your opinion about approaching some of them (patients or doctors) and asking some questions?

Exercise 5.10 (Gregory et al. 2013) The objective of this exercise is to gain some experiment in interviewing someone and taking notes.

before Make an appointment about 'mobility' (or another topic that matches your academic or professional interests) with a person that you know (friend, family member, or co-worker). Prepare a set of questions to ask the interviewee.

during During one hour, conduct an interview. The topic is mobility, which the interviewed person may interpret in a variety of forms and for which you can decide how to ask the questions. Develop the interview as a conversation, using the answers to conduct it in a natural way. Listening and understanding the perspective of the person are fundamental for the success of the interview. Don't record the interview, but take notes of the most important terms and sentences.

after Describe (3–4 pages) the interview, including: your name; a pseudonym of the interviewee, sex, age and occupation; a short description of the scenario in which the interview took place. Include your pre-prepared questions and describe the conversation. Conclude with your thoughts about the interview: the interactions and the dynamic that was established with the interviewed person, her analysis of the addressed topics, and any other observations that may seem pertinent.

Exercise 5.11 (Gregory et al. 2013) The objective of this exercise is to experiment observing the behaviour of the persons in a given public place and to take notes.

before Choose a public place that seems interesting to observe. Examples of possible places are: airport, train station, car park, hospital waiting room, post office, bank, supermarket, canteen, bar, gymnasium, museum, library. The local should allow you to observe and take notes without being disturbed.

during For a time span of 60 min, observe and register the movements, interactions, sounds, space and everything that captures your attention. Take short notes while you observe. If someone asks you what you are doing, tell her that you are doing a school project.

after Elaborate on the material that you have collected. You are expected to write at least three pages. Justify the place you selected and present the respective blueprint. Describe what you have observed (indicating who, where, when, how) and provide your own interpretation. Conclude the text with your interpretation of a rule that is applicable to the local. What patterns were you able to identify? what exceptions were detected? are there persons that behave distinctly from the others?

Chapter 6
Requirements Negotiation and Prioritisation

Abstract The quality of a software application is highly dependent on its ability to meet the necessities of customers and users, so it is crucial to capture and specify all requirements that the system must possess. However, many projects have more candidate requirements than those that can be built without exceeding the available time and budget. This chapter presents and discusses the need to allow stakeholders to collectively negotiate the requirements in order to decide which ones shall be incorporated in the system. The chapter also discusses the importance of assigning priorities to requirements, in the context of a system project. The aspects related to requirements prioritisation and some of the most popular prioritisation techniques are also presented.

6.1 Requirements Negotiation

When there is no consensus among the stakeholders with respect to the requirements, a *conflict* arises amongst the contradicting views, as well as among the stakeholders that proposed those conflicting requirements. For example, in the context of software development, a user can request the application to be rebooted, in case an error is detected, while a different user may suggest that the application continues its operation. Another example of a conflict is when the quality of service demanded by the users is in contradiction with the financial restrictions indicated by the client that pays for the application.

Techniques for assigning priorities, in order to solve conflicts, can be generically divided in two main classes: prioritisation methods and negotiation approaches. The former are based on assigning values to different aspects of the requirements, while the latter are focused on attaching priorities to the requirements after the agreement of the stakeholders. The processes of conflict resolution are quite complex, due to the natural incompatibilities and divergencies among the stakeholders.

Conflicts are inevitable and are present in all human relationships and in all societies. Negotiation is a solution for solving conflicts (Illustration 6.1). One negotiates when there are at least two alternatives to choose from. For the interested parties, those alternatives present common and conflicting interests. Many engineering projects

© Springer International Publishing Switzerland 2016   119
J.M. Fernandes and R.J. Machado, *Requirements in Engineering Projects*,
Lecture Notes in Management and Industrial Engineering,
DOI 10.1007/978-3-319-18597-2_6

Illustration 6.1   The negotiation as a technique for the resolution of conflicts

fail, due to the lack of negotiation of the conflicts in the requirements. The estab-
lishment of agreements amongst the stakeholders is a quite difficult mission, since it
demands skills that most requirements engineers had never explored in a systematic
way.

   Negotiation is, in its essence, a process of decision-making carried out in a context
of strategic interaction or interdependency. More simply, **negotiation** is a basic social
process that is used with the objective of solving conflicts. The negotiation implies
a minimum of two participants, whose decisions are mutually contingent. In the

negotiations, the participants resort to information and power, to try to influence the behaviour of the remaining participants. The execution of a negotiation, in the context of a conflict among requirements, seeks to identify one of a set of acceptable and technologically-viable alternatives.

The conditions that transform the negotiation into a fundamental mechanism for solving conflicts are the following, according to Lewicki and Litterer (1985):

- there exists a conflict of interests among two or more parties;
- there are no rules, methods or referees to solve the conflict;
- it is impossible to opt for various alternative solutions;
- the contract cannot be undone, which requires a compromise regarding the collective interest to be reached.

6.1.1 Negotiation Process

The negotiation process, as Fig. 6.1 illustrates, essentially goes through three steps: (1) pre-negotiation, (2) negotiation, and (3) post-negotiation.

Pre-negotiation, as the name suggests, happens before the negotiation and is focused on the definition of the negotiation problem, identification of the stakeholders, elicitation of their objectives, and analysis of the objectives to detect the conflicts. For each identified conflict, there is a range of alternatives, one of which should be accepted by the negotiators. In this step, one should clearly identify what is the problem, in order to define the objective of the negotiation. Next, the stakeholders must be identified. Actually, identifying the right persons in the negotiation processes has the same challenges as doing it for any other process. It is not a trivial task, but it is essential for the success of the process. Subsequently, the stakeholders must indicate what are their objectives for the system under consideration. The collected objectives are examined in order to identify the conflicts.

In the *negotiation* step, the stakeholders are effectively involved so that they can negotiate and reach an agreement about the issues in conflict. The objective must be always to reach mutually-beneficial solutions that can be acceptable for all parties. Thus, one should look for alternatives to solve the problem, by exchanging offers and counter-offers or proposing solutions of mutual gain. Based on the available alternatives, the stakeholders try to agree with the "best" one. To obtain the agreement of the parties, it is necessary to establish evaluation criteria, to compare the value of the different alternatives. To this end, it can be convenient to organise a preparatory session to reach a consent about these criteria.

Fig. 6.1 Main steps of the negotiation process

Table 6.1 Comparison between consensus and majority

Consensus	Majority
Promotes cooperation	Promotes competition
All stakeholders have the same decision-making power	The decision-making power is equivalent to hierarchy and popularity
Stakeholders either agree or disagree	Stakeholders vote against or in favour or abstain themselves
All ideas are discussed and explored	The majority decides which ideas to consider

A negotiation can be resolved in two ways: (1) *consensus* or unanimity, which requires the acceptance from all implicated parties, or (2) *majority*, in which prevails the wish of the most numerous group. Each one of these ways has advantages and disadvantages as presented in Table 6.1.

> "Consensus is the process of abandoning all beliefs, principles, values, and policies in search of something in which no one believes, but to which no one objects."
>
> *Margaret Thatcher (1925–2013), politician*

When some parties are not satisfied with the final result, win-lose situations appear. This world view between winners and losers is deeply rooted in north-american society and is reflected in many aspects of ordinary life. For example, in collective sports in which the objective is to make more points than the opponent, many european sports (for instance, football, handball, rugby, field hockey) allow ties, something that usually cannot happen in North-American sports (for instance, american football, basketball, ice hockey, baseball). Hence, games that finish in a draw are extended until the tie is broken and consequently a winner is found. Boehm et al. (2001) present some situations of win-lose situations common in software engineering, considering three parties: client, users and engineer (see Table 6.2). These situations must be avoided, despite the fact that they can look attractive for the benefited parties,

Table 6.2 Common win-lose situations in software engineering; adapted from (Boehm et al. 2001)

Solution	Winners	Loser
Rapidly create a cheap product, but with low quality	Engineer, client	Users
Add useful functionalities, but with low relevance	Engineer, users	Client
Obtain a price much lower than the market value	Client, users	Engineer

since they have little support in a medium-term time horizon. In some cases, they can contribute to comprising or even destroying the good existing relationship.

It is evident that reaching a consensus is preferable to majority rule, but it is more difficult to obtain. However, the requirements engineer must strive to reach consensual solutions for conflicts, since they are more likely to result in win-win situations in which all can win. A classical example where a win-win situation could bring advantages for all is when two persons are disputing an orange. After some discussions they conclude that the solution is to cut the orange in two halves, one for each person. Surprisingly, it was found that one of the persons ate the pulp and threw away the peel, while the other did exactly the opposite, since he made an orange peel jam and discarded the pulp. The identification of the interests of each person could have maximised their gains, with one getting all the pulp and the other all the peel.

> "Most people think of success and failure as opposites, but they both are products of the same process."
>
> *Roger von Oech (1948–), creativity theorist*

In *post-negotiation*, the result of the negotiation is analysed and evaluated, suggesting if necessary and possible a renegotiation or trying to find better solutions. In this latter case, one can find an alternative that increases the satisfaction of one party, without decreasing the other's. In this step, it is also necessary to ensure that the stakeholders commitment with respect to the agreement reached is kept over time.

6.1.2 Postures and Strategies

Negotiation processes involve various aspects that must be taken into consideration. Next, the different postures that a participant can have in face of a negotiation and the strategies that can be followed are presented and discussed. Generically, there are five postures that can be taken in a conflict situation: inaction, competition, accommodation, collaboration, and compromise (Thomas 1976). This classification, as shown in Fig. 6.2 is based on two axes: the desire and willingness that each individual, implicated in a negotiation process, has in satisfying (1) the other interests and (2) his interests. In a negotiation process, it is important to identify in which typical posture each participant can be framed, since that affects how the person that coordinates the process and the other participants behave.

Each one of those five postures are next summarised:

- **Inaction** means indifference towards the result of a negotiation.
- **Competition** presupposes an emphasis that favours one's own interests at the expense of the others', which often leads to win-lose situations.
- **Accommodation** involves the satisfaction of the others' interests, without considering his own. This may mean that one of the parties is willing to favour the others, because he simply wants the negotiation to be finished or to ensure that the

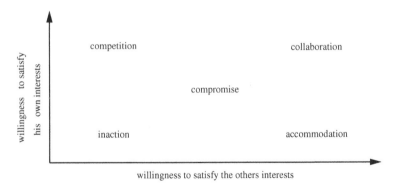

Fig. 6.2 Typical postures in conflict situations throughout a negotiation

other parties are satisfied (to claim something in the future), or because the issues
are much more relevant to the other parties.

- **Collaboration** is focused on the satisfaction of the concerns of all the implicated
  parties, to find alternatives that can satisfy all the stakeholders. The emphasis is
  on looking for win-win situations.
- **Compromise** involves concessions to find a reasonable and balanced solution that
  leaves all parties satisfied.

> "Never compete with someone who has nothing to lose."
> *Baltasar Gracián (1601–1658), philosopher*

Concerning various strategies, there are two principal ones that can be followed
in a negotiation: (1) integrative and (2) distributive. In an *integrative strategy* (or
symbiotic), it is assumed that the parties are willing to cooperate in searching for
consensual solutions. The aim is to reach an agreement in an imaginative and collaborative way and achieve win-win situations, resorting, without any restriction, to
shared information. In this strategy one tries to satisfy the interests of each party,
as completely and fairly as possible. It is common to verify that one of the parties
concedes to the interests of the other in an explicit strategy of assumed submission,
with the purpose of preserving a positive relationship in the future. Alternatively, the
parties can try hard to maintain a good relationship and increase the gains for both
of them.

The *distributive strategy* (or predatory) is based on the assumption of firm and
inflexible postures, where the participants are only concerned with their needs and are
indifferent to the others'. Thus, it is frequent that aggressive and intimidating behaviours (for example, mockery, attack, or intimidation) are adopted. The negotiation
becomes a game, where one either wins or loses. The success for each stakeholder
depends on his capacity to make concessions in a gradual way, exaggerate in the
value of the concessions that he proposes, minimise the value of the concessions of

Table 6.3 Characteristics of negotiation strategies

Integrative	Distributive	Principled
Participants are friends	Participants are opponents	Participants solve problems
Objective is agreement	Objective is victory	Objective is a rational result, achieved amicably
Make concessions to promote the relation	Claim concessions as a condition for the relation	Separate the persons from the problem
Change your position easily	Insist on your position	Focus on interests not in positions
Find a consensus solution	Force your solution	Create various possible solutions to choose later
Insist on consensus	Insist on your opinion	Insist on the use of objective criteria

the other parties, and hide information. In this strategy, the egos of the participants tend to emerge and often the positions taken usually put in the background the true necessities or interests.

In the middle of the two extreme strategies, an intermediate strategy can be considered. The so-called *principled strategy* tries to establish an equilibrium among the advantages and disadvantages of each one those two strategies (Fisher et al. 1999, p. 11). In this strategy, the participants must follow four fundamental principles: (1) work as a team, tackling the problem instead of attacking themselves; (2) focus on interests not positions; (3) create several options to later choose one; (4) use, if possible, objective criteria to compare the options. Table 6.3 summarises some of the characteristics of the three negotiation strategies previously discussed.

"Let us never negotiate out of fear. But let us never fear to negotiate."
John F. Kennedy (1917–1963), politician

6.2 Requirements Prioritisation

In the project of a system, it is necessary to establish the set of candidate requirements, i.e., the requirements that are amenable to be incorporated in the system, due to their relevance from the onset. Based on this set, the subset that includes the most important requirements must be selected. Requirements *prioritisation* is a technique that aids in identifying those fundamental requirements and that can be viewed as the process that sorts a set of requirements, according to various criteria (Illustration 6.2).

Illustration 6.2 Prioritisation allows stakeholders to contribute to the definition of the importance of the requirements in a given project

Some users find worthless the need to prioritise, since they argue that all the candidate requirements will be incorporated. However, if a development team is not able to deliver all the requirements in the agreed deadline, the stakeholders must consent with respect to the subset to be implemented in the first place. In software products, the market pressure frequently implies very aggressive commercial strategies that propose launching those products, as soon as possible, in good enough versions. Many technological startups adopt the *minimum viable product* (MVP) strategy that suggests launching in the market products with the fundamental set of requirements and nothing more (Ries 2011, p. 93). In these cases, the product does not include obviously all the candidate requirements, so it is necessary to use prioritisation techniques to distinguish the most valuable requirements from the ones less interesting. The prioritisation process supports the following tasks:

- to decide about the fundamental requirements of the system;
- to establish an order for the requirements to be implemented;
- to implement a part of the requirements, but still be able to construct a product that satisfies and pleases the users;
- to compare the benefit of each requirement with its cost;
- to estimate the satisfaction of the client with respect to the product;
- to handle conflicting requirements and to negotiate them with the stakeholders.

Whenever the expectations of the client are high, deadlines short, and resources reduced, it is necessary to make sure that the system includes the most essential functionalities. Defining the (relative) importance of each functionality permits one

Fig. 6.3 Main steps of the
prioritisation process

to serialise the construction of the system to obtain the highest value at the smallest
cost. The users and the development team must cooperate in the requirements prioritisation. The team does not always knows which characteristics are most valued
by the clients and additionally the clients often are not able to evaluate the cost and
the technical difficulty associated with each requirement.

According to Karlsson et al. (1998), the prioritisation process consists of three
steps, as illustrated in Fig. 6.3:

- **Preparation**: In this step, the requirements are structured in accordance with the
  principles of the prioritisation technique in use. A set of stakeholders is selected,
  providing them all the necessary information about the requirements and the prioritisation technique itself.
- **Execution**: In this step, the stakeholders conduct the requirements prioritisation
  based on the information that was previously provided. The criteria must be agreed
  upon before initiating the prioritisation.
- **Presentation**: This step presents the results to the stakeholders implicated in the
  process. Some prioritisation techniques require calculations that must be executed
  before presenting the results.

The challenge in the execution of a prioritisation process lies essentially in selecting, from a given set of candidate requirements, those that make sure that all the main
interests, all the restrictions and all the preferences of the different stakeholders are
taken into account and that the system market value is maximised. Since the knowledge that the analysts have about the requirements varies throughout the process, the
assignment of priorities must be realised in an iterative way, with different abstraction
levels and based on distinct informations, during the system lifecycle.

In dynamic domains with enormous competition, where environmental changes
happen constantly, the requirements prioritisation and negotiation must be frequently
repeated. Agile methods, discussed in Sect. 2.4.2, are adequate for producing software in that type of contexts and naturally put great emphasis on the prioritisation,
a task that may be repeated in each development cycle.

"There is nothing permanent except change."
Heraclitus of Ephesus (c. 535 B.C.–c. 475 B.C.), philosopher

6.2.1 Criteria and Scales

Requirements can be assigned priorities according to different criteria. When a single criterion is used to assign priorities to the requirements, it is trivial to completely sort them. If more than one criterion is used, then the difficulty of sorting increases, since it is necessary, in these cases, to weight and balance those various criteria. It is not always clear how to establish that order. In particular, it is well known that the introduction of cost, as a criterion to consider, typically implies the decrease of the overall priority of some requirements, if their incorporation in the system is expensive.

Various criteria can be considered in the prioritisation: importance, urgency, usefulness, penalty, (in)satisfaction, (development) time, cost, volatility, and risk. However, it is not practical to consider an excessive number of criteria. Those that should effectively be considered in a given project must be carefully chosen taking into account the specific situation.

Priorities according to a given criterion can be assigned with different measures and scales. There are four main types scales of measure: nominal, ordinal, interval and ration (Aczel 1995, p. 4–5).

In a *nominal scale*, the numbers or labels serve to categorise the objects, that can be either equal or different between them. This scale serves just to identify an items belongs or not to a category. Hence, this scale cannot be used for prioritisation purposes, since it does not allow the requirements to be sorted. Examples of this type of scale are postal codes, civil statuses, and part numbers.

An *ordinal scale* permits seeing if an object is more important that another one, but not by how much. The values of the scale indicate the relative position and how far the objects are between them, according to the considered criterion. The data are differentiated and sorted by values expressed in a scale with an arbitrary origin. Examples of this scale are the Likert scales that allow one to answer questions with fixed alternatives (for instance, I totally agree, I agree, neutral, I disagree, I totally disagree).

The *interval scale* permits one to quantify the distances among the measurements, but it does not present a natural null point. It is thus possible to assign a meaning to the difference between the numbers, but not to their ratio. A classical example is the temperature scale (Celsius or Fahrenheit), where one cannot assume the zero point (point of nullity or null point) or say, for instance, that a given temperature is the double of another.

In a *proportional scale* (or ratio scale), it is not only possible to quantify the differences between two measurements, but also to ensure certain advantageous mathematical conditions, due to the existence of the null point. In particular, the quotient of two measurements is meaningful, independently of the measurement unit. Examples of ratio scales are age, salary, price, sales volume and distance.

6.2.2 Techniques

Briefly, the essential objective of any prioritisation technique is to associate a value to each considered requirement, so that it is possible to establish a (total or partial) order among them. Next, some of those techniques are presented, starting from the most simple ones and ending with the most complex ones.

Top-10

In this technique, each stakeholder must select, from the set of candidate requirements, those ten that he considers as the most important, without however establishing any order among them. This technique is appropriate when the number of stakeholders is high, especially when all of them have a similar importance and relevance for the project under consideration. The major challenge in the utilisation of this technique lies in deciding which requirements must be considered, based on the various top-10 lists that are available. This is the reason why those lists are not internally ordered, since that could result in situations in which a stakeholder would see his sixth requirement considered, in contrast with another stakeholder whose first requirement would be considered. Even so, it is difficult to make sure that situations similar to this do not happen, since it is almost inevitable to choose for each different user a distinct number of her top-10 requirements for incorporation in the system. More complicated to justify are the cases in which no requirement from the top-10 of a given user is considered.

Ranking

With ranking, an ordinal scale is used and the requirements are totally ordered, that is, ties are not allowed. The most important requirement stays in the 1st position, the second most important requirement in the 2nd position, and so on until the least important requirement is placed in the last position. In this technique, each requirement has a unique position, contrarily to what happens in grouping, but it is not possible to observe the relative difference between the ranked requirements, as happens in AHP and the 100-unit test (next discussed). Ranking seems to work well when there is only one person using it, which occurs, for example, in some agile methods. However, its application becomes more complex when it is necessary to combine the ranking tables from various persons with distinct perspectives. It is possible to combine many tables in just one, calculating for instance the average of the positions of each requirement. However, this solution may originate a final list in which some requirements are tied, forcing the adoption of some rules to break the ties that this technique does not allow by construction.

Grouping

Grouping, also referred to as numerical assignment, is one of the most used prioritisation technique and consists in distributing the requirements into different groups. Although the number of groups can vary from case to case, it is typical to use three groups (for instance, critical, desired, and optional; or expected, normal, and fascinating). For most cases, that tripartite division seems to be sufficient and a lot of authors suggest it, for example, Pfleeger and Atlee (2009, p. 152) and Pressman (2009, pp. 131–132). However, the MoSCoW method (IIBA 2007, p. 102) recommends four groups, adding a fourth option (W) that, in practice, postpones the definition of the group: M (*must*; requirements that must be considered), S (*should*; requirements that should be considered), C (*could*; requirements that are desirable but not necessary), W (*won't*; requirements that will not be implemented in a given version, but that may be considered in the future).

When applying this technique, one must avoid that the users tend to classify a great majority of the requirements as critical, since unbalanced distributions, like 85 %–10 %–5 %, are frequent. A solution is to fix minimum quotes to each group (for instance, 20 %), but that can mean that the usefulness of the priorities diminish, since users can be forced to put requirements in groups where they genuinely were not interested to. Another obvious limitation of this technique is that only a partial ordering of the requirements is obtained, since inside the same group the requirements are not prioritised.

100-unit test

This technique consists in requesting each stakeholder to distribute 100 units (for example, points or dollars) among the candidate requirements (Berander and Andrews 2005). It presents some limitations, especially when the number of requirements is high; if there are, for example, 50 requirements, there exist on average only two units for distributing to each one. This implies that there is little room for manoeuvre to assign priorities to the requirements and to precisely differentiate them. A solution for this limitation is to consider a bigger number of units (for instance, 10,000), which permits a larger precision in the differentiation among the requirements. It is also advisable to use a tool (a simple spreadsheet is sufficient) to make sure that the units make exactly a total of 100.

In principle, it should not be possible for a person to redo the test, after knowing the global result. If that situation was possible, that person could put all the points in her favourite requirement, if it was not at a top position. This situation could grossly bias the final result. Alternatively, that person could remove her points, with respect to a requirement that had received many units from other stakeholders, as long as the inclusion of that requirement in the system would not be affected. Thus, she could maintain that requirement and promote other requirements to higher positions, assigning them the removed units.

Analytical hierarchy process

The AHP (analytical hierarchy process) (Saaty 1980) is a systematic method to cope with complex decision-making problems, that has been adopted in different domains. It is used for solving multi-criteria decision problems, that is, to take decisions in scenarios in which various factors have different relative importances. Its utilisation, in requirements engineering contexts, is based on the comparison of the importance of all pairs of requirements, using a scale from 1 to 9. Table 6.4 shows for various comparisons made verbally the respective numerical values that this method proposes.

The comparison between all the n requirements of a system can be represented in a bi-dimensional matrix A, whose values obey the rules that are next described ($1 \leq i; j \leq n; i \neq j$). The entries on the main diagonal are always filled in with the value 1, since each one of those requirements is *equally important* to itself

Table 6.4 Association between the verbal comparisons and the numerical values for the AHP method

Verbal comparison	Value
Equally important	1
Moderately more important	3
More important	5
Much more important	7
Extremely more important	9
Intermediate values	2, 4, 6, 8

($a_{i,i} = 1$). Then, the remaining matrix entries outside the diagonal are filled in, based on the judgement and the intensity of the importance according to Table 6.4. When entry $a_{i,j}$ is filled in with the value X, the diametrically opposed entry ($a_{j,i}$) can be automatically filled in with the value $1/X$.

> "The human brain is built to compare; it's Darwinian to consider an alternative when one presents itself."
>
> *Helen E. Fisher (1945–), anthropologist*

Please consider the following matrix, which refers to five requirements (R_1 to R_5) and includes their respective pair-wise comparisons. It is filled in in accordance with the rules previously indicated. For example, the value 7 in line R_4 and column R_1 ($a_{4,1}$) represents the perception/opinion of the person who is making the comparison that requirement R_4 is *much more important* than requirement R_1. The value in the diametrically opposed entry ($a_{4,1}$) is $1/7$, that is, the reciprocal of 7.

$$
\begin{array}{c}
\begin{array}{ccccc} R_1 & R_2 & R_3 & R_4 & R_5 \end{array} \\
\begin{array}{c} R_1 \\ R_2 \\ R_3 \\ R_4 \\ R_5 \end{array}
\left[\begin{array}{ccccc}
1 & 2 & 2 & 1/7 & 1/2 \\
1/2 & 1 & 2 & 1/5 & 1/2 \\
1/2 & 1/2 & 1 & 1/5 & 1 \\
7 & 5 & 5 & 1 & 5 \\
2 & 2 & 1 & 1/5 & 1
\end{array} \right]
\end{array}
$$

After filling in the matrix, which necessarily must be accomplished by a person, the AHP method must execute a series of steps that can be totally automated. Firstly, the values in the columns are summed up, obtaining the result shown in the last line of the matrix:

$$
\begin{array}{c}
\begin{array}{ccccc} R_1 & R_2 & R_3 & R_4 & R_5 \end{array} \\
\begin{array}{c} R_1 \\ R_2 \\ R_3 \\ R_4 \\ R_5 \end{array}
\left[\begin{array}{ccccc}
1 & 2 & 2 & 1/7 & 1/2 \\
1/2 & 1 & 2 & 1/5 & 1/2 \\
1/2 & 1/2 & 1 & 1/5 & 1 \\
7 & 5 & 5 & 1 & 5 \\
2 & 2 & 1 & 1/5 & 1
\end{array} \right] \\
\begin{array}{cccccc} \Sigma & 11 & 10.5 & 11 & 1.74 & 8 \end{array}
\end{array}
$$

The next step consists in the normalisation of the columns, making sure that the total of each column is 1. To obtain that result, one must divide the value of each entry of that column by the respective total (calculated in the previous step). For the example, the following matrix[1] is obtained:

$$
\begin{array}{c}
\quad\ R_1 \quad\ R_2 \quad\ R_3 \quad\ R_4 \quad\ R_5 \\
\begin{array}{c} R_1 \\ R_2 \\ R_3 \\ R_4 \\ R_5 \\ \Sigma \end{array}
\begin{bmatrix}
0.091 & 0.190 & 0.182 & 0.082 & 0.063 \\
0.045 & 0.095 & 0.182 & 0.115 & 0.063 \\
0.045 & 0.048 & 0.091 & 0.115 & 0.125 \\
0.636 & 0.476 & 0.455 & 0.575 & 0.625 \\
0.182 & 0.190 & 0.091 & 0.115 & 0.125 \\
\end{bmatrix} \\
\quad\ 1 \qquad 1 \qquad 1 \qquad 1 \qquad 1
\end{array}
$$

In the next step, the sum of the values in each line is calculated, thus obtaining the relative valuation of each requirement. For the example under discussion, the result of this step is shown in the last column of the following matrix:

$$
\begin{array}{c}
\quad\ R_1 \quad\ R_2 \quad\ R_3 \quad\ R_4 \quad\ R_5 \quad\ \Sigma \\
\begin{array}{c} R_1 \\ R_2 \\ R_3 \\ R_4 \\ R_5 \\ \Sigma \end{array}
\begin{bmatrix}
0.091 & 0.190 & 0.182 & 0.082 & 0.063 \\
0.045 & 0.095 & 0.182 & 0.115 & 0.063 \\
0.045 & 0.048 & 0.091 & 0.115 & 0.125 \\
0.636 & 0.476 & 0.455 & 0.575 & 0.625 \\
0.182 & 0.190 & 0.091 & 0.115 & 0.125 \\
\end{bmatrix}
\begin{array}{c} \mathbf{0.608} \\ \mathbf{0.500} \\ \mathbf{0.424} \\ \mathbf{2.767} \\ \mathbf{0.703} \end{array} \\
\quad\ 1 \qquad 1 \qquad 1 \qquad 1 \qquad 1 \qquad 5
\end{array}
$$

Based on the values of the last column, one can sort the five requirements in accordance to their importance: R_4 (2.767) > R_5 (0.703) > R_1 (0.608) > R_2 (0.500) > R_3 (0.424).

The use of redundancy in the pairwise comparisons allows judgement errors to be detected and also a consistency ratio to be calculated. For example, it may sound impossible that A has higher priority than B, B higher priority than C, but A lower priority than C. In the example previously considered, one verifies that R_5 is more important that R_1, R_1 is more important that R_3, but R_5 is equally important as R_3. However, situations of this type may appear in reality, since it is useful to consider a certain degree of ambiguity in the comparisons when tens of requirements must be handled. In a system with n requirements, $\frac{n \cdot (n-1)}{2}$ comparisons are necessary, which means that their number has a quadratic growth in relation to the number of requirements. For example, if there are 50 requirements, 1 225 comparisons are required; If the number of requirements grows to 70, then 2 415 comparisons are needed, almost the double. This fact makes this technique adequate only for cases that involve a relatively low number of requirements.

---

[1] The total in some columns apparently is not exactly equal to 1, but this results from the fact that the values were rounded to 3 decimal digits.

   The ranking that results from AHP may be strongly affected by modifications, even if small, in the set of candidate requirements or in the evaluations resulting from the perceptions of the stakeholders. The possibility of the occurrence of some of these changes are at the origin of the controversy installed in the scientific community and that is based on two opposing arguments. On one side are the researchers, for instance (Holder 1990), that defend that such possibility highlight intrinsic theoretical problems of the AHP method. On the other side are the ones, for example (Forman and Gass 2001), that argue that this occurs due to the flexibility of the method and that this characteristic constitutes one of its strongest advantages.

   There are some approaches, based on AHP, that were proposed specifically for prioritising software requirements, being probably the most popular the cost-value approach presented by Karlsson and Ryan (1997).

Choice of technique

As a suggestion, one should use the most simple prioritisation technique that is expected to be successful, allowing thus decisions to be taken with the best cost-benefit ratio. The most sophisticated techniques must be reserved for the most sensitive situations that require more careful analysis. In many cases, the availability of automatic tools can ease the choice of those sophisticated techniques.

   Taking into account some of the limitations mentioned before, the combined use of prioritisation techniques is often needed. A possibility consists in using both grouping and ranking techniques; the requirements are firstly divided in groups with different priorities and then classified, that is, ordered, inside each group. The requirements triage is another combined technique that is matched by what is done in hospitals to classify, in calamity situations or war scenarios, the wooden persons that present themselves in the emergency services. The medical staff divides the victims in three groups: (1) those that will die, even if assisted; (2) those that will recover, even if not assisted; and (3) those for whom a medical assistance may have a significant impact. Analogously, when one assigns priorities to requirements, there are those that must be in the system (critical), those that the system does not need to consider (optional) and those that require more treatment (desirable). Hence, the requirements that are part of the last group must be handled by another technique (for instance, AHP, 100-unit test, ranking).

6.3 Summary

Conflicts arise among stakeholders when there is no consensus about requirements. Conflict resolution is generally a complex process, due to disagreements among various stakeholders. It can be accomplished with negotiation approaches, which define the priorities of the requirements after agreement from the stakeholders, or prioritisation methods, that seek to assign values to different requirements aspects.

   A negotiation can be solved by consensus, in which it is necessary the acceptance by all parts, or by the majority, in which the will of the largest group prevails.

The chapter discusses also the different postures that a participant can have when negotiating (inaction, competition, accommodation, collaboration and compromise) and the strategies that can be followed (integrative, distributive, principled).

Requirements prioritisation is a technique that aids in the identification of the fundamental candidate requirements. It can be seen as a process that sorts a set of requirements, helping to distinguish the most valuable requirements from those that have less interest. Some prioritisation techniques (top-10, ranking, grouping, 100-unit test, AHP) are presented.

Further Reading

There are many books about negotiation in general, a subject that is quite fascinating and that helps in addressing technical problems, but also daily ones, with strategies that have proved useful. A classic book is authored by Fisher et al. (1999). Another book Fisher and Shapiro (2005), by the same principal author, deserves to be read, since it adds the emotion to the negotiation process. A very complete book, written by Lewicki et al. (2010), is also a good reference.

The literature that addresses negotiation in the more specific context of the requirements is not abundant. The more relevant material is the chapter written by Grünbacher and Seyff (2005). An analysis of the practical utility of the different scales in the context of the requirements prioritisation can be found in Karlsson et al. (2006). Some tools also exist to support the negotiation of requirements, for example, Boehm et al. (2001).

Interesting and useful is also an incursion into game theory, whose application in negotiation contexts is often highly pertinent. A well-presented textbook on classical game theory is Peters (2008), that provides a rigorous coverage of the subject. The book by Bazerman and Moore (2012) is worth reading, especially Chap. 10.

In addition to AHP, there are other multi-criteria decision algorithms that can be adopted for supporting the prioritisation process. A good source for studying some of those algorithms is the book organised by Figueira et al. (2005).

Exercises

Exercise 6.1 (Naveda and Seidman 2006, pp. 71–72) While eliciting requirements, the analyst has registered, for each requirement, the name of the person that firstly proposed that requirement and the date in which that proposal was made. Which of the following missing pieces of information will have a bigger impact (negative, due to its absence) on the activities associated with change management?
(a) Traceability, (b) Requirements type, (c) Priority, (d) Source.

Exercise 6.2 (Naveda and Seidman 2006, pp. 37–38) Which of the following sentences better portrays the requirements management process?

1. Measuring the requirements quality permits saying that they remain unchanged over time.
2. A requirement, once rejected, should remain rejected.
3. Change is inescapable, so the requirements management process should take into consideration that fact.
4. One should use a computer-based tool to trace a set of (functional and nonfunctional) requirements.

Exercise 6.3 Consider that you have six requirements (R_1 to R_6). Rank them, according to the importance (weight 70%) and user satisfaction (weight 30%) criteria, with the AHP method.

importance

	R_1	R_2	R_3	R_4	R_5	R_6
R_1	1	2	2	1/5	1/2	1
R_2	1/2	1	2	1/5	1/2	1/4
R_3	1/2	1/2	1	1/5	1	1/4
R_4	5	5	5	1	5	2
R_5	2	2	1	1/5	1	2
R_6	1	4	4	1/2	1/2	1

user satisfaction

	R_1	R_2	R_3	R_4	R_5	R_6
R_1	1	1	3	4	1/2	1/4
R_2	1	1	2	5	1/3	1/4
R_3	1/3	1/2	1	1/2	1/7	1/7
R_4	1/4	1/5	2	1	1/7	1/9
R_5	2	3	7	7	1	1/2
R_6	4	4	7	9	2	1

Exercise 6.4 Which of the following criteria/dimensions should exist in a mechanism for the classification of requirements negotiation processes?
(a) Support to renegotiation, (b) Strategy to resolve conflicts, (c) Level of automatisation, (d) Support to documentation.

Exercise 6.5 Describe each possible posture that a stakeholder can have with respect to the negotiation process.

Exercise 6.6 coordinated by the instructor: Divide the participants in the session in pairs for playing arm wrestling (also known as *bras de fer*). If you are reading this book outside the context of an academic course, find a friend or family member to play with you. There are 30 s to play and the objective for each player is to maximise the number of points. At the end, each participant indicates the number of points obtained, the winner being obviously the one that was able to accumulate more points.

Exercise 6.7 (Raiffa 1982, pp. 262–267) coordinated by the instructor: Divide the students in groups of three persons, designated by A, B and C. If you are reading this book outside the context of an academic course, find two friends or family members to play. The objective of the game for each participant is, during at most 30 min, to negotiate within her group an agreement that maximises her individual gain, taking into account the point limits defined in the following table:

Agreement	Persons	Total gain
1	Just A or just B or just C	0
2	A and B	118
3	A and C	84
4	B and C	50
5	A and B and C	121

Before the negotiation starts, each participant should carefully analyse this table, establish a strategy, and write it for further analysis. In this phase, the participants do not communicate among them. During the negotiation, if two persons wish to have a private conversation, they can do it just once, during 2 min. At the end of the negotiation, the group must indicate which persons are in the agreement and how (i.e., with numbers) the gain is divided among them. Those that do not participate in the agreement have zero points. When all negotiations are concluded, the results for the three different roles (A, B and C) in each group are announced. Based on that information, each group analyses the performance of each element and discusses what happened in the negotiation.

Exercise 6.8 coordinated by the instructor: Imagine that a given city is threatened by a bombing attack. There are 12 persons interested in seeking protection in an air-raid shelter, that however can only accommodate six persons:

men

- A 40 year old violinist, addicted in cocaine;
- A 25 year old lawyer;
- A priest with 75 year old;
- A 20 year old atheist, responsible for several murders;
- A 28 year old physicist, that only enters the shelter if carrying a gun;
- A 21 year old poet, that adores to declaim his poems;
- A 47 year old homosexual;

women

- The wife of the lawyer, that is coming out of the madhouse; the lawyer and the wife prefer to stay together, even if outside the shelter, than to be separated;
- A 34 year old prostitute;
- A university student that did a vow of chastity;
- A 12-year-old girl with a low IQ;
- A 32 year old woman with mental disabilities, that has epilepsy.

1. Make your list, deciding which six persons shall enter the shelter.
2. Form a group with three other persons and decide upon a unique list with the six persons to enter the shelter, base on the individual lists.
3. All the students should now produce a list of six persons, based on the lists of each group.

Chapter 7
Writing in a Natural Language

Abstract This chapter is focused on describing a set of practical recommendations to write requirements in good English (e.g., write simple sentences, use a limited vocabulary, and avoid ambiguity) and on analysing the structure of a document template for writing requirements. It is expected that the reader realises on the advantages that result in having requirements written in a clear, methodic and standardised way. The chapter also discusses ambiguity in natural languages, emphasising the special care that must be taken in the use of such type of languages to write requirements. Although some of the contents in this chapter can be used in other natural languages, the target is the English language.

7.1 Guidelines for Writing

Within the development of a system, it is necessary to communicate with the various stakeholders, which may explain why the writing of the requirements in a natural language is almost inescapable. It is not expectable that all the stakeholders are able to interpret, for example, requirements specifications in formal or rigorous formats, in the mathematical sense of those terms. The software engineer must be capable of writing requirements in a natural language. Moreover, the engineer, in general, must know how to communicate either orally or in a written format with other engineers, as well as with any common person. Writing in an effective way is however a task prone to errors and problems. What one writes will not always be interpreted in the desired form, which explains why writing requirements, according to a set of principles and recommendations, is important. Writing requirements is not specially difficult, but requires continuous enhancement, through training and practice.

There are no magical formulas to write well, since writing is an art. The option for writing requirements in a natural language without any restriction, i.e., in a totally free format, presents several advantages. There are no limits to expressiveness, it is supposedly understandable by all literate persons, and it does not require any specific training. However, free writing presents normally many ambiguities and lack of rigour and eventually the final quality varies from case to case. Hence, it is crucial to find some principles that aid in writing good requirements. Obviously, there is not a

© Springer International Publishing Switzerland 2016

J.M. Fernandes and R.J. Machado, *Requirements in Engineering Projects*,
Lecture Notes in Management and Industrial Engineering,
DOI 10.1007/978-3-319-18597-2_7

unique correct way for writing requirements. In this section, some recommendations are suggested, which, if followed, allow software requirements to be consistently written, creating a minimal reference for homogenisation (Illustration 7.1).

7.1.1 Issues to Consider

This subsection presents some issues that, when followed, aid in writing requirements in a more adequate form, making the process more systematic and eliminating some problems. This first set presents constructive recommendations that must be considered when writing, while the second one identifies forms of writing that must be avoided or banned.

> "Before the interest in writing, there's another one: the interest in reading. And things are going badly when you only think in the former, if previously the taste for the latter was not consolidated. Without reading no one writes."
>
> *José Saramago (1922–2010), novelist*

Illustration 7.1 Requirements written in a natural language must follow a set of recommendations in order to ensure that the documents are homogeneous in terms of content and style

Basic writing skills

As a starting point, it is important to master the grammatical rules, namely in what concerns orthography and punctuation. Writing ability can be refined, by reading and essentially by practicing. To achieve a higher level in the quality of what is written, it is necessary to write a lot and often. It is important that one understands that the perfect requirement does not exist, so it is difficult to decide if a given requirement has already reached an acceptable written form. Additionally, it is likely that many requirements can be modified as the stakeholders re-think what they actually need (i.e., what their real necessities are). Thus, it is natural that the writing process is iterative and that the requirements are improved and modified through the development.

Technical writing

Writing requirements for engineering systems must obey the basic rules of technical writing. We present some of these rules below. The language must be simple, clear, and precise. Figures of speech, like metaphors or similes, should not be used, since the words should be adopted in their denotative meanings, with no space for possible alternative interpretations. Technical writing must be done in an impersonal, objective, clear, humble, and polite style.

> "The first quality of style is clarity."
>
> *Aristotle (384 B.C.–322 B.C.), philosopher*

Standard formats

Requirements writing must follow a standardised format, to give coherence and uniformity to all documents, easing thus the interpretation of the texts. There are different patterns for the user requirements and system requirements, due to the different purposes of each requirements type (see Sect. 3.4).

Generically, the following format for sentences that represent user requirements is suggested (Alexander and Stevens 2002, p. 67):

- a subject that indicates the type of **users** that benefit from that requirement;
- an intended result to achieve with the requirement (using a predicate):
  - verb (i.e., **functionality** to be performed),
  - **description**, using other sentence elements to complete the predicate: (direct, indirect, prepositional) objects, predicatives, and adjuncts.
- a mechanism to allow a **test** for the requirement to be defined.

An example of a user requirement that follows this format is:

(users)	The hotel receptionist
(functionality)	should visualise
(object/concept)	the room number of a guest,
(test)	2 s after making the request.

Some agile methods propose the adoption of user stories, another format for writing requirements. A *user story* is a simple and short description of a functionality, made in the perspective of the person that needs it. A user story is materialised through a set of sentences, written in common or business language, that describe what the users do or need within the scope of their functions. User stories normally have the following format:

As a <type of user>, **I want to** <objective> **for** <reason>.

Using the example of the requirement related to the hotel receptionist, its writing as a user story and according to this format is:

As a hotel receptionist, **I want to** visualise the room number of a guest **to** call him if someone wants to contact him.

This form of writing requirements puts the stakeholder as the focus of attention and eases the identification of the sources. Due to the form in which it is written, it is almost certain that this requirement was requested by a receptionist. It is hard to believe that this requirement was solicited by the client or by the hotel manager.

With respect to system requirements, the standard format that is proposed for writing them is the following:

- A subject, either the **system** under development or a design **entity** that is related to the requirement.
- an intended result to achieve with the requirement (using a predicate):
  - verb (i.e., **functionality** to bring about),
  - **description**, using other sentence elements to complete the predicate: (direct, indirect, prepositional) objects, predicatives, and adjuncts.

An example of a system requirement that follows this format is:

(system/entity)	The signal of the battery
(functionality)	must turn on,
(description)	when the charge is lower than 20 mA h.

The writing of non-functional requirements can also adopt a specific format, that we propose to be the following:

- the **system** under development or a design entity that is related to the requirement.
- a **quality** to be achieved with the requirement:

  - verb (in some cases, a verbal form of "to be" or "to have", or something equivalent);
  - object (i.e., a description).

All the examples of non-functional requirements given in Sect. 3.3 follow this format. Four of those examples are here reproduced just for reference:

The product shall be easy to use for illiterate persons.
The product must continue to function at 30 m under water.
The product must be prepared to be translated to any language.
The source code of the product programs should contain comments.

Short and simple sentences

Each requirement must be represented by one sentence and each sentence should just represent one requirement. The objective is to have the requirements written clearly and this presupposes short and simple sentences. The sentences must be affirmative and written in the active voice. Therefore, one must avoid negative sentences or written in the passive voice. Sentences must also be self-contained, so that it is not necessary to read other requirements or documents to fully interpret it. In particular, references to other documents should be limited, especially those with restricted and limited access.

Sometimes, there is a tendency to write sentences in a verbose style, since one thinks that they are complete and cover all the relevant situations. This is rarely true, being more common that those sentences are tedious and very hard to interpret.

> "Make everything as simple as possible, but not simpler."
> *Albert Einstein (1879–1955), physicist*

Conciseness is mandatory and if it is possible to write a requirement with, let us say, seven words, one should not write it with ten or twelve, to not make the readers unnecessarily waste time. However, writing short sentences is not easy and paradoxically it normally takes (much) more time to write a short sentence than a long one. Shortening a sentence requires the text to be analysed and rethought so that it sticks to the essential, a task that takes longer than writing the original sentence.

A sentence must be as short as possible, as long as its clarity is not affected. Finding the equilibrium between size and clarity is something that hardly is obtained at the first attempt. It is thus necessary to (re)write the requirement several times, trying several alternatives and distinct words, until an adequate formulation is obtained. A good rule of thumb is to avoid the same word appearing two or more times in the same requirement. In principle, it should be possible to find another form for the text without repeating so many times the same word.

The following example can be considered a requirement written in a too long format:

> It is a device that allows the client to record and store his favourite programming to watch whenever he wishes, in addition to pause, go forward and backward. Additionally, it allows different programs to be recorded and viewed at the same time, so the client will always be able to continue to watch a program that for some reason he could not finish at that moment.

The next formulation, simpler and shorter, is decidedly preferable and seems, in its essence, to transmit the same message:

> The client, with the device, watches and records different programs at the same time.

Limited vocabulary

The utilisation of a limited vocabulary must constitute a concern of all involved in writing requirements. It is also important to avoid the utilisation of terms that may create confusion, especially synonyms of words that represent important concepts. Natural languages, like English, are very rich in their diversity. For instance, there are many ways to see: contemplate, detect, inspect, look, note, observe, regard, spot, view, watch, and witness. For example, if the users of a system are designated by the term 'student', that designation should be used coherently throughout the document. One should not fall into the tendency, very common in many natural languages, of using synonyms to make the text less repetitive. This idea may make sense in many literary texts, but not in technical documentation. For the student example, terms like 'apprentice', 'disciple', 'graduate', 'learner', 'pupil', 'scholar', or 'undergraduate' should not be used. A reader, when confronted with a new word, might be uncertain if it refers to the same concept or to a new one, a situation that one wants obviously to avoid.

"A synonym is a word you use when you can't spell the other one."
Baltasar Gracián (1601–1658), philosopher

Acronyms and abbreviations must be used with great care, although in some cases it is difficult not to resort to expressions already established and in common use. It is more convenient and simple to use terms from the computing area, like ERP, FEM, GNU, JPEG, MP3, PDF, RAM, SMTP, rather than the expanded forms. Equally, well established abbreviations from other domains can be used: AIDS, DNA, Laser, LED, NBA, NBC, Radar, Sonar, TNT, TV, UNICEF, and USA. In any case, one should avoid creating new acronyms if their purpose is only to facilitate the writing process. The practical consequence is that reading becomes more difficult.

7.1.2 Issues to Avoid

Next, some aspects that must be avoided when writing requirements are presented. The suggestions that are here discussed must be complemented with the recommendations discussed in the previous subsection to ensure that the requirements are written in a systematised, organised, and coherent way.

Ambiguity

Ambiguity is an important question to handle when writing requirements. Its occurrence means that there are two or more possible interpretations for a sentence. The situations of ambiguity must be corrected, with the objective of making a sentence clearer. If the sentence was not written by the person that detected the ambiguity, it may be necessary to contact its author, since often she is the only person able to eliminate the ambiguities from the sentence. When adequate, the sentences can be complemented with other materials (tables, figures, schemes), to make the meaning clearer. Due to the importance of ambiguity, Sect. 7.3 is entirely devoted to it.

Ambiguity is manifested also when two or more conflicting requirements are defined, a situation that must obviously be rectified. That conflict may presuppose a contradiction between the interests of distinct persons, being thus recommendable to solve it through negotiation techniques (see Sect. 6.1).

Vague terminology

Many words that are used informally to indicate desirable characteristics for a given system are too vague, so they should not be used in the context of a requirements document. Some words or expressions that should be avoided are: easy to use/learn, versatile, flexible, intuitive, modern, improved, efficient, approximately, more possible, minimal impact. In practice, when one uses this type of expressions, the meaning is not clear and, as a consequence, it is impossible to verify if the requirements are or not satisfied.

The following example presents a non-functional requirement that, despite being clear, uses vague terms that should be banned:

> The e-mail application must have an intuitive user interface.

The difficulty here, related to a non-functional requirement, lies in the fact that it is hard to decide when an interface is *intuitive*. A solution is to complement the writing of the requirement with the definition of verifiable criteria. Thus, a possible form of verifying if the e-mail application incorporates or not the requirement consists in performing a series of experiments, with real persons that represent the different user profiles. One may define, for instance, that the interface is considered as intuitive if, in the first utilisation of the application, with no help, 85 % of the users are able to perform the most basic tasks after 25 min. It is necessary to identify what are those basic tasks. In the example, they could be, for instance, configuring the application with the user account details, receiving, replying, composing and sending messages. The verification of the users performance can be made by live observation or, in alternative, through recording the application being used for further visualisation and analysis. It should also be established what is the threshold (i.e., the percentage of users that are able to perform the basic tasks in the stipulated time) that defines if the graphical interface is or is not intuitive.

Illusions and fantasies

In engineering, there are no perfect components. One should thus avoid any sort of wishful thinking, in which one is trying to reach the impossible. Hence, it is imperative that the project team has a realistic and rational attitude, rejecting any requirement that may be associated with an illusory or unrealistic idea. In that sense, one should remove any requirement that include expressions like: 100 % reliable, totally safe, never fails, satisfies all users, handle all unforeseen situations. The three following examples show requirements that do not follow this guideline:

> The printer shall be always operational.
> The software application shall handle all unexpected errors.
> The biometric authentication system shall be 100 % reliable.

"When we do fantasy, we must not lose sight of reality."
                                                    Walt Disney (1901–1966), cartoonist

Multiple requirements

Requirements that contain coordinating conjunctions are especially susceptible to create ambiguity situations (see Sect. 7.3). For that reason, it is a good practice to avoid the use of coordinators like: for, and, nor, but, or, yet, and so (also referred to as FANBOYS). The solution in these cases consists in dividing one requirement into simpler ones. Consider the following requirement that contains the 'or' term:

> The guest shall pay the bill with money or credit card.

It is preferable to divide this requirement into two simpler ones:

> The guest shall pay the bill with money.
> The guest shall pay the bill with credit card.

A second example that should be avoided is the following:

> The system shall produce a beep or visual signal to be sent to the director or secretary.

It is difficult, in this case, to decide what actually must be implemented to fulfil this requirement. Assuming that the term 'or' is inclusive, Table 7.1 summarises the nine alternatives that exist for this requirement, which shows that its interpretation is not unique. There are also two other possible interpretations, although rather unlikely, that consist in sending each one of the signals in an exclusive way to each person: either sending the beep to the director and the visual signal to the secretary, or vice-versa.

Design

One should avoid indicating how the system will be able to satisfy a given requirement. This normally implies that the requirement has too much detail and that design decisions are taken prematurely. The requirement is thus over-specified, which limits the solution space of the system and may mean that one is losing the opportunity to consider more adequate technical solutions.

Table 7.1 Alternatives for satisfying an ambiguous requirement that includes two occurrences of the term 'or'

Beep	•	•	•				•	•	•
Visual signal				•	•	•	•	•	•
Director	•		•	•		•	•		•
Secretary		•	•		•	•		•	•

The focus of the writing process should be on functionalities that the system will provide to its users. The way the system will be constructed so that those functionalities are incorporated should not be a concern of those responsible for defining and writing the requirements. There are several symptoms that can be searched to know if one is not complying with this recommendation. The inclusion in the requirements of component names, materials, database fields, or technological aspects should be avoided:

> The clock shall present the current civil local time, by receiving code time signals from a radio station.
> The guest shall complain about the hotel services through a form available in the web.

The indication on how to receive the current civil local time is a design element that limits prematurely the engineering options. For instance, the local time can be synchronised directly from an atomic clock, from the internet, or manually. Indicating that the complaint about the hotel services must be done through a web form can also reduce the alternatives. Will it be more adequate that the complaint is made through an email message sent to a known address? Is there any consumer protection law that requires the complaint to be formalised in a book that exists for that purpose, thus eliminating the need for a web form? Without an answer to questions like this and unless there are many strong reasons for these technological choices, the requirements must be written as follows:

> The clock shall present the current civil local time.
> The guest shall complain about the hotel services.

Speculations

Requirements should be obtained from the identified sources and, in particular, from the stakeholders. Hence, the requirements are things that the stakeholders request and not that the requirements engineer decides that they will need them. Requirements that the users did not request, but that the project team believes to be useful or interesting, should not be considered. This recommendation, especially relevant for tailor-made software systems, rests on the premise that most probably those additional requirements will: (1) not be paid by the client, (2) not be valued by the users, (3) delay the delivery of the system, or (4) be incorporated in the system at the expense of excluding requirements requested by the stakeholders.

Suggestions

Generally, a suggestion represents a requirement that a stakeholder finds interesting, but that is not fundamental to fulfil his necessities. It is often viewed as creating a "by-the-way" opportunity, that is, it represents something that, if included in the system, represents an additional, but collateral, characteristic.

Suggestions of requirements are expressed by terms like: can, could, should, may be, probably. Requirements written with these words must be ignored or classified with low priority levels, since there are certainly other requirements with higher priority that deserve the attention of the development team. The following example represents a requirement that expresses a suggestion or possibility that must be rewritten in a more assertive way, if it is really an important requirement, or removed if it is a suggestion:

The product should probably be able to receive signals from the outside, even when used in a train moving at least at 130 km/h.

This type of suggestions are often verbalised in other forms. The stakeholder could start the sentence with "it would be interesting if the product could ...". The basic idea is however to understand if the requirement is indeed important or just supplementary.

Project plans

Project plans and the way the project is scheduled are important aspects to consider within the scope of any project, but should not be included in the requirements document. Hence, one should avoid the inclusion of dates, phases and project activities in the requirements. That type of information must be available in a different document (project plan).

7.2 Template for the Requirements Document

The project team is responsible for the edition of the requirements document. It is expected that the team puts all its professionalism in this task. It is an activity that demands a great care, namely with respect to all the details of the document, like style, presentation, and structure. The definition of a generic template for the documents that specify the requirements is an important aspect, since there is a great diversity of engineering systems and projects. Without that template, the degrees of freedom are excessive, making the documents quite different from case to case. The existence of a pre-defined structure is very useful, specially for highly-complex systems, to help to efficiently manage the writing process.

The advocates of agile methods suggest that one should not put excessive effort in documentation tasks, since the requirements vary a lot, making it difficult to keep the documents updated. Instead of resorting to formal documentation, the requirements are registered as user stories, in small cards. However, registering the requirements in a written format is necessary, even if that is done in a lighter form than the one proposed by the most traditional development approaches.

"Voluminous documentation is part of the problem, not part of the solution."
Tom DeMarco (1940–), software engineer

In this book, the generic structure for the requirements documents suggested by Robertson and Robertson (2006, Chap. 10 and Appendix B) is adopted. This structure should be understood as adaptable to the context of an engineering project, which means that sections can be added or eliminated, the latter being more likely to happen than the former. For example, if there are no performance-related non-functional requirements, the respective section can be eliminated. A software producer can also create its own version, according to the domains in which it operates, the type of projects in which it is involved, or the clients with whom it cooperates. In any case, for each specific project, it is recommendable to identify who will be reading the documents, since based on their profiles, the contents may be different. The template for the requirements documents, in a first sectioning level, has the structure that Table 7.2 illustrates. Next, each one of the 27 sections of the template are summarily described. The interested reader should consult the book by Robertson and Robertson (2006), if more details are needed.

1. Purpose of the system

This section of the document must describe the business context in which the system will be inserted and indicate the reason that was in the origin of its development. It must also indicate, in one or two sentences, what are the objectives of the project, i.e., what are the reasons why the client wishes to have the system.

2. Client, customer, stakeholders

Here one indicates the name of the client, since he is the one paying for the system development and thus he must be satisfied when it is deployed. The name of the clients must also be indicated in the case of a tailor-made software system and the potential customers must be characterised, in the case of software products. Finally, the roles and job titles must also be specified, as well as the names of the persons that have some interest with respect to the system. The various types of stakeholders, indicated in Sect. 5.2, should be here considered.

"The customer is always right."
Harry Gordon Selfridge (1858–1947), retail magnate

3. Users of the system

The potential users of the system are listed in this part. One must also indicate, for each type of users, the respective level of importance or involvement, as discussed in p. 85. Whenever possible, the participation (either related to the scope or to the duration) that is expected for each type of users should be mentioned during the

Table 7.2 Generic table of
contents for the requirements
documents; adapted from
Robertson and Robertson
(2006, Appendix B)

Project triggers	
1	Purpose of the system
2	Client, customer, stakeholders
3	Users of the system
Restrictions of the project	
4	Mandatory restrictions
5	Taxonomy and definitions
6	Facts and assumptions
Functional requirements	
7	Scope of the work
8	Scope of the system
9	Functional and data requirements
Non-functional requirements	
10	Appearance
11	Usability
12	Performance
13	Operational
14	Maintenance and support
15	Security
16	Cultural and political
17	Legal
Project issues	
18	Open issues
19	Immediate solutions
20	New problems
21	Tasks
22	Migration to the new system
23	Risks
24	Costs
25	User manual
26	Waiting room
27	Ideas for solutions

requirements elicitation. If that participation is defined from the very beginning, the users are aware of what is expected from them. Otherwise, the expectations of the parties risk to be distinct.

4. Mandatory restrictions

In some scenarios, there are constraints on how the system must be constructed. If this is the case, in this template section one should indicate the solutions or technologies to be adopted, as well as the reasons that led to them. One should also specify the implementation environment in which the system will operate. Here, it is relevant to identify which systems will interact with the system under consideration, both upstream and downstream. The environment in which the users of the system operate must also be described. If the system is to be used, for example, inside libraries, it is expected that it does not emit sounds or noise. If it is to be used inside a boat, one is expecting that it is water- or at least moisture-resistant. Budget and time constraints (dates and deadlines) should also be expressed here. Often, all these restrictions must be negotiated, since the client may be asking, for a very low price, too many requirements for the available time.

5. Taxonomy and definitions

This section must include a glossary with the meaning of all the terms and acronyms used in the document. Here, new nomenclature should not be invented, being important to adopt the one that is used in the problem domain. This section is very important, since it is with words that humans communicate and, inevitably, some words gain different meanings when the domain is different. The definitions of the terms must be balanced, neither too extensive, nor too simple. In this section, all the information flows and storage elements used in the models must be defined. In particular, the informations related to the context diagram must be specified.

6. Facts and assumptions

Here the assumptions made by the development team, related to any aspect that can affect the system, are recorded. Please consider the following examples of assumptions that can be included in this part: laws and political decisions, operation environment, dependencies with respect to persons and other systems, components available for implementation. It is important to register these assumptions as early as possible to avoid misunderstandings among those that participate in the project. These assumptions are transitory, since, when the requirements documentation is closed, each one will be either a requirement or a restriction. For example, if an implementation component is indeed available then it is a technological restriction to use. Otherwise, the functionality of that component constitutes a requirement that the development team must implement.

7. Scope of the work

This section includes an analysis of the current situation of the business processes, namely the manual and automated processes that can be replaced or altered by the system at hand. Sometimes, this type of analysis was already realised, as part of the feasibility study that conducted to the system development. Based on this analysis, it is possible to understand the consequences that result from introducing the system and thus opt for the most favourable alternatives. It is convenient also to include the context diagram (that can be obtained by filtering the use case diagram), that identifies the scope of the system to be built.

8. Scope of the system

In this section, the use case diagrams of the system can be included, identifying thus the frontier between the system and the actors. The textual descriptions of the use case must also be included, possibly considering also the scenarios for each use case. More detailed informations about use cases can be found in Sect. 8.4.2.

9. Functional and data requirements

Each requirement must be specified according to the following fields:

- unique identifier: numerical counter, or of other type, used for referencing univocally the requirements;
- type of requirement: numerical indication of the document section where it is included (9 functional, 10–18 non-functional, 4 restrictions, 18–27 issue);
- use cases: lists of use cases that require of the requirement;
- description: sentence that describes the functionality offered by the requirement;
- source: indication of the origin of the requirement (for example, name of the person or law);
- satisfaction of the client: level of satisfaction, if the requirement is included in the system;
- dissatisfaction of the client: level of dissatisfaction, if the requirement is not included in the system;
- priority: level of priority of the requirement;
- conflicts: identification of other requirements that are in conflict with this one;
- dates: register of the dates in which the requirement was created and changed.

10. Appearance

This section is related to the non-functional requirements of appearance and style of the system. A discussion about this kind of requirements is made in Sect. 3.3.1.

11. Usability

This section includes the non-functional requirements related to the easiness of use, personalisation, internationalisation, easiness of learning, comprehensibility, and accessibility. This type of requirements is discussed in Sect. 3.3.2.

12. Performance

This section covers the non-functional requirements related to the processing time of the tasks, response times, accuracy of the results, reliability, availability, robustness, fault-tolerance, storage capacity, scalability. This type of requirements is analysed in Sect. 3.3.3.

13. Operational

The non-functional requirements related to the (environmental or technological) context in which the system will operate are expressed in this section. Sect. 3.3.4 addresses this kind of requirements.

14. Maintenance and support

In this section, one registers the non-functional requirements related to maintenance, repair, support (p. ex, through a call centre), and adaptability. A discussion about this type of requirements is made in Sect. 3.3.5.

> "A man should keep his friendship in constant repair."
> *Samuel Johnson (1709–1784), poet*

15. Security

This section addresses the non-functional requirements related to the confidentiality, integrity, and privacy. Sect. 3.3.6 presents this type of requirements.

16. Cultural and political

This section contains the non-functional requirements related to cultural and political aspects. In Sect. 3.3.7, this requirements of this category are discussed.

17. Legal

This section handles the legal requirements, that is, the conformity of the system with the established laws. A discussion about this kind of requirements is available in Sect. 3.3.8.

18. Open issues

Here one registers all the issues that were addressed in the analysis phase, but could not be solved. The issues to include must have some relevance for the project. One can indicate, for instance, topics related to possible changes in the commercial strategy of the hardware producers or the possibility of the appearance of new laws that might affect the system under development.

19. Immediate solutions

In this part of the document, one should list existing systems that can satisfy the requirements of the stakeholders. In some cases, it may be preferable to buy one of these systems rather than build a new one from scratch. One should also indicate (libraries of) software components, whose use in the project is evaluated as relevant. Finally, it is relevant to list systems that can be copied and modified. The so-called free or open-source software systems fit here, because their respective licenses enable free access to code for various purposes (use, sharing, copy, distribution and modification).

20. New problems

The objective of this section is to anticipate the detection of potential conflicts that may result from operating the system, so that adequate measures can be taken before those conflicts actually occur. One should describe how the system will affect and influence the current working environment. In special, it is necessary to evaluate if the system will imply some restructuring of the organisation or new forms of work. The way the system under development will interface with existing systems in the environment can also be included. The human component is also relevant, namely to understand how the various stakeholders will react to the new system. This area can reveal to be very critical. It is one of the questions that is handled by socio-technical approaches that try to increase the success of the use of software systems in contexts strongly affected by human, social, and organisational factors.

21. Tasks

Here one describes the details of the approach to be followed for developing the system. This section is particularly relevant when the development process established in the context of the software producer has to be modified due to special circumstances associated with the project. If possible, an estimation about the time and resources needed to carry out the project tasks should be included.

22. Migration to the new system

This section identifies the tasks that allow the migration of the current system to the new one. It is important to characterise here, for example, how the new system will be installed, which data conversion mechanisms are necessary, which support tasks is necessary to operationalise, and which formalities are required to remove the old system from service. If no current system exists, then this section should be eliminated.

23. Risks

The potentially adverse circumstances that can have a negative impact on the quality of the final application are called risks. They include two important characteristics: (1) probability of occurrence, since it is not sure that the risk will happen, (2) consequences of occurrence, since if the risk becomes real, there are undesired consequences. In this section, one should specify which risks that may affect the project are the most probable and critical. The probability and impact of each risk must also be estimated.

"The biggest risk is not taking any risk... In a world that is changing really quickly, the only strategy that is guaranteed to fail is not taking risks."
Mark Zuckerberg (1984–), cofounder of Facebook

24. Costs

The cost of a requirement is associated with the effort to incorporate it in the system. Based on a set of specified requirements, it is possible to estimate the cost of the system, using a costing estimation model. The best known model to estimate the cost of software is most probably COCOMO II (Boehm et al. 2000). It is important that the client is informed at this stage about the estimated cost of the system development. This information is typically conveyed in a global way, indicating a price for the system as a whole, but it may be interesting or useful to present the cost of each individual requirement.

25. User manual

This section presents the list of documents the user is provided as part of the system. Depending on the type of system, one may have to create tutorials, user manuals, complete or quick reference manuals, installation manuals, service manuals, and emergency procedures manuals. One should also identify who will be responsible for creating the documentation. When writing a manual, one should bear in mind the purpose of the document, its level of detail, who will read it, and who will update it. This part should also address the training about the system, identifying who is responsible for preparing and giving the training sessions.

26. Waiting room

This section serves for registering requirements that will not be incorporated in the current version of the system, but that can potentially be included in a future version. The candidate requirements which ended up not being considered, are not deleted from the document, but rather transferred to this section. Thus, the effort spent in the elicitation, elaboration, negotiation, and writing of these requirements is not wasted, and may be reused in the future. The requirements included here may be prioritised with some of the recommendations proposed in Sect. 6.2.

27. Ideas for solutions

The central objective of requirements elicitation consists in identifying the user necessities, without addressing the technical solutions that can support them. However, it is natural that some innovative ideas to solve the problem might arise. This part of the document serves for registering those ideas, which thus remain separated from the requirements. The format to use here is not fixed, but it is important that it permits one to get back to the registered ideas. In many projects, this section may not be necessary. If this is the case, it should be removed from the document.

7.3 Ambiguity

Whenever one writes a requirements document, the principal objective is not highlighting the writing abilities of the author. Instead, the aim in general with a technical document is fundamentally clarity. In this sense, it is important to avoid ambiguous sentences, since they cause difficulties and doubts in the interpretation. The characteristics of the words, expressions, or sentences that express more than one possible understanding is called **ambiguity**. It is widely used in poetry, publicity or jokes, but it must be banned from technical texts.

According to Berry (2008), there are basically three approaches to avoid the ambiguity of natural languages:

1. write less ambiguously;
2. detect ambiguity either manually, or with the help of tools;
3. use a restricted natural language so that it becomes unambiguous (but also less natural).

This book does not handle exhaustively, neither too theoretically, the topic of ambiguity. The main aim is to make the reader aware of some aspects that need to be considered, when she is writing requirements in a technical context, so that ambiguous sentences are not incorporated in the documents. Recognising the existence of ambiguity situations during the writing process is quite important. This awareness reveals the attention related to the need in identifying those situations, allowing thus more information to be captured with the objective of rewriting the texts to make them clearer.

The attentive reader has noticed that previously the verb 'to avoid' was used, since even someone that is tuned to ambiguity, cannot detect all the situations in which such phenomenon is manifested. One assumes here that ambiguity is not introduced on purpose by those that wrote the requirements, but as a result of carelessness, negligence or lack of sensitivity for the topic. Next, ambiguous situations that can happen in the English language and that should be avoided are analysed. It is also relevant to consider that often a requirements document is written in a language by non-native persons. This is always problematic issue since it means that the text gets more unclear and ambiguous.

Homonymy and polysemy constitute possible causes of lexical ambiguity. The distinction between these two concepts has been highly discussed by the experts in linguistics. A situation of *homonymy* occurs when two or more distinct and unrelated meanings accidentally share the same lexical form. An example is the word 'left' that can be the past tense of the verb 'to leave' and the opposite of right. *Polysemy* occurs when the same lexical unit supports two or more distinct meanings, but somehow semantically related. An example is the word 'coffee', that can mean (1) the fruit of a plant or (2) a drink made from that fruit. In practical terms, the difference between homonymy and polysemy is not always easy to establish, but it seems also not very relevant for the objective that is here discussed. What is relevant is the various meanings that are associated with those terms.

"False words are not only evil in themselves, but they infect the soul with evil."
Socrates (470/469 B.C.–399 B.C.), philosopher

In the following example, the word 'secretary' can mean a person or a piece of furniture:

> The secretary of the director is tall.

In some cases, the context of the sentence is sufficient to eliminate ambiguity, since, as the following example shows, the sympathy or its lack are characteristics that can be associated with humans, but not to objects:

> The secretary of the director is unsympathetic.

Another common cause of ambiguity is related to the use of conjunctions. Consider a first example:

> The managers inform the directors and the secretaries, because they are responsible for editing the document.

In this case, it is not totally clear if the persons responsible for the edition of the document are only the secretaries or if the directors are also included. It is not easy to disambiguate this sentence, since the context permits both readings to be perfectly valid. In fact, the directors seem also to have responsibility for editing the document. There are two solutions for writing the sentence without that possible double understanding:

> The managers inform the directors and the secretaries, because the latter are responsible for editing the document.
> The managers inform the directors and the secretaries, because both of them are responsible for editing the document.

This example shows the problem of interpretation that exists when, after an enumeration, one writes something that can be applied to all the elements enumerated or just to the last one. The enumerations are responsible for various ambiguity situations, so they need to be handled with especial attention.

In the common language, the terms 'and' and 'or' have often identical practical effects, which implies that their use must be well analysed. Both are conjunctions and in the case of 'or' its value can be either inclusive or exclusive. This situation differs from what happens in mathematics, where the Boolean operator AND (gives the value one if and only if all the inputs are one; otherwise gives the value of zero) is distinct from the Boolean operator OR (gives the value one if at least one input has a value of one; otherwise gives the value of zero). A term that should not be employed in technical documentation is 'and/or', since it is a source of ambiguity. It puts two conjunctions at the same point of the text, which induces more reading alternatives. It is always preferable to use only 'or' than 'and/or'.

In the next pages, sentences where the terms 'and' and 'or' may have similar effects in terms of interpretation. In the first couple of sentences, the interpretation is the same, regardless of using 'and' or 'or':

> The club accepts, as members, men <u>and</u> women.
> The club accepts, as members, men <u>or</u> women.

The lack of a second interpretation results from the fact that it is impossible for a single person to be simultaneously a man and a woman. In the next example, however, the two alternatives are not always exclusive:

> The club accepts, as members, tall <u>and</u> thin men.
> The club accepts, as members, tall <u>or</u> thin men.

The first sentence has two possible interpretations. One indicates that the club accepts as a member any man that is simultaneously tall and thin. In this case, 'and' is used as an element that unites two adjectives that qualify the same subject. Another interpretation, in which 'and' is used as an element that adds sets, indicates that the club accepts tall men and also thin men. The second sentence seems to have just this last interpretation.

Ambiguity seems to disappear if one writes the sentences with some repetition or using singular subjects.

> The club accepts, as members, tall men <u>and</u> thin men.
> The club accepts, as members, any tall <u>and</u> thin man.

When plural nouns, namely nouns linked by 'and', constitute the subject of a sentence, it can be unclear whether the entities constituting the subject are to be considered individually or collectively. The next three sentences clarify this aspect:

> Anthony and Barbara shall call Charles.
> Anthony and Barbara shall each call Charles.
> Anthony and Barbara, acting collectively, shall call Charles.

Whenever 'and' and 'or' are used in sentences related to prohibitions or permissions, their use is also similar, as the following example shows:

> It is forbidden to cut flowers <u>and</u> step on the grass.
> It is forbidden to cut flowers <u>or</u> step on the grass.

In this case, what makes sense is both things to be prohibited, i.e., none of them is allowed. The first sentence could be interpreted as being prohibited simultaneously both, but not one of them individually. Thus, (1) one would be allowed to cut flowers as long as he does not step on the grass, and (2) one would be allowed to step on the grass, as long as she does not cut flowers. However, from the context of the sentence, which characterises a situation of prohibition, one realises that the only interpretation that makes sense is to not permitted none of the situations.

> "There is no greater impediment to the advancement of knowledge than the ambiguity of words."
>
> *Thomas Reid (1710–1796), philosopher*

A more correct (i.e., less ambiguous) form of writing these sentences is: 'It is forbidden to cut flowers and it is forbidden to step on the grass'. The problem of this sentence is that it is not economical, forcing the expression 'it is forbidden' to be written twice.

In lists of items it may be the case that the 'and' (or 'or') used for separating the two last elements of the list can confuse the reader. The following example tries to highlight this problem:

> This law is valid in Angola, São Tomé and Principe and Mozambique.

If one does not know the names of this Portuguese-speaking African countries, one is not sure about which of the two following possible lists is the one:

- (1) Angola, (2) São Tomé and Principe and (3) Mozambique.
- (1) Angola, (2) São Tomé and (3) Principe and Mozambique.

One solution to eliminate this ambiguity is to enumerate the countries, as is indicated in the first alternative show above. Another is to use the symbol '&' in 'São Tomé & Príncipe'. If the order of the items is not relevant, São Tomé and Príncipe can be placed in the first place: 'São Tomé and Príncipe, Angola and Mozambique'. It is also possible to put each alternative in quotation marks: 'Angola', 'São Tomé e Príncipe' and 'Mozambique'. It is also possible to separate the countries by ';': 'Angola; São Tomé and Príncipe; and Mozambique'. In english, this problem is eliminated if one uses the so-called Oxford comma (also called serial or Harvard comma), i.e., a comma used before the conjunction in the lists with three or more alternatives. Thus, the list with the three countries ('Angola, São Tomé and Príncipe, and Mozambique') has always a unique understanding. The Oxford comma may however introduce ambiguity as the next example shows:

Barbara calls her father, Anthony, and Charles.

Here, the Oxford comma creates ambiguity about the Barbara's father because the punctuation is identical to that used for an appositive phrase, leaving it unclear whether this is a list of three persons (1, her father; 2, Anthony; and 3, Charles) or only two (1, her father, who is Anthony; and 2, Charles). So, the use of the Oxford comma must be analysed from case to case.

Some forms of ambiguity are related to the use of possessive pronouns in the third person (in the singular or plural), after having referred to more that one subject. Consider the following example:

The director calls the manager about <u>his</u> problems.

This is a classic situation of ambiguity that occurs in the English language. In this case, the possessive adjective 'his' can be related to either the director or the manager (since both are male) and the context of the sentence does not help in solving that ambiguity (Illustration 7.2). If the two persons have different sexes, this problem does not exist:

Anthony calls Barbara about <u>his</u> problems.
Anthony calls Barbara about <u>her</u> problems.

A similar ambiguity exists in sentence like the following one, where it is not clear the father of whom (Anthony or Anthony's friend) works at a hotel.

The director calls the doctor
about his problems

Illustration 7.2 The ambiguity of the English language requires special care, in particular when writing requirements for engineering projects

Anthony's friend and his son work at a hotel.

Possible, but more verbose, solutions are the following:

Anthony's friend and his friend's son work at a hotel.
Anthony's friend work at a hotel, as well as Anthony's son.

It is common that texts of juridic nature to be ambiguous, due to the style in which they are written. Consider the following example, that does not follow the recommendation to write short sentences:

The XYZ company will be liable for any damage resulting from the negligent exercise of this mandate, namely as a result of the non-compliance by the XYZ company of any established legislation, or as a result of producing capital losses with respect to the value of the financial assets at the date on which they were entrusted to the XYZ company to be managed within this agreement contract, and the fiscal tax obligations of the incomes obtained as a result of the said mandate.

> "Be brief, for no discourse can please when too long."
> *Miguel de Cervantes (1547–1616), novelist*

In this case, it is relevant the fact the following part of the sentence is between commas: 'namely as a result of the non-compliance by the XYZ company of any established legislation'. Hence, a possible interpretation is that the XYZ company is liable for any damage that from the negligent exercise of the mandate (being explained, between commas, in what consists that negligent exercise) or as a result of producing capital losses. The 'production of capital losses' is not, in this interpretation, included in the description of what is considered as a negligent exercise. If that is the message being conveyed, a way to eliminate this ambiguity is to adopt the following sentence:

The XYZ company will be liable

1. for any damage
   a. resulting from the negligent exercise of this mandate, namely as a result of the non-compliance by the XYZ company of any established legislation, or
   b. as a result of producing capital losses with respect to the value of the financial assets at the date on which they were entrusted to the XYZ company to be managed within this agreement contract, and
2. the fiscal tax obligations of the incomes obtained as a result of the said mandate.

However, it is possible to interpret that the description about the negligent exercise covers also the production of capital losses. In this case, the society is liable for any damage that results from the negligent exercise of the mandate. That negligent exercise can, namely, result from the non-compliance of any established legislation or the production of capital losses. In this case, a form of eliminating the ambiguity is opting for the following formulation:

The XYZ company will be liable

1. for any damage resulting from the negligent exercise of this mandate, namely
   a. as a result of the non-compliance by the XYZ company of any established legislation, or
   b. as a result of producing capital losses with respect to the value of the financial assets at the date on which they were entrusted to the XYZ company to be managed within this agreement contract, and
2. the fiscal tax obligations of the incomes obtained as a result of the said mandate.

Another distinct form of formulation results in breaking the sentence in two shorter ones, as is next shown:

The XYZ company will be liable for any damage resulting from the negligent exercise of this mandate and the fiscal tax obligations of the incomes obtained as a result of the said mandate.

One understands as negligent exercise the non-compliance by the XYZ company of any established legislation, or the production of capital losses with respect to the value of the financial assets at the date on which they were entrusted to the XYZ company to be managed within this agreement contract.

7.4 Summary

The use of requirements written in a natural language is common in engineering projects, because it fosters communication among the various stakeholders. Writing is a task prone to errors and problems, since what is written often is interpreted in a different way than intended.

The chapter focuses on the description of a set of practical recommendations for writing good requirements in English (for instance, use standard formats, write short sentences, use a limited vocabulary, avoid ambiguity and vague terms). Having the requirements written in a clear, methodic, and standardised way results in advantages for all participants in a given project, because it increases the effectiveness of the communication.

This chapter also discusses the structure of a standard template for documenting requirements, which includes 27 possible sections. The definition of this generic template for the requirements documents is a relevant aspect, since the natures of the systems and projects that are addressed by engineering are very diverse. The suggested template is especially useful in highly complex projects.

Ambiguity-related issues in natural languages are also discussed. Ambiguity happens when there are two or more possible interpretations for a sentence, making thus the text unclear. Those situations must obviously be corrected, to eliminate difficulties and doubts of interpretation. The topic of ambiguity is addressed with the aim of alerting the reader to some aspects to be taken into account when writing requirements in an engineering project, so that ambiguous sentences are not incorporated in the documents.

Further Reading

Readers who wish to improve their writing skills have many bibliographic sources. Here, we recommend the work by Larocque (2003) that presents practical guidelines and rules (such as, keep sentences short, and keep to one main idea per sentence; use the right word, prefer active verbs and the active voice, cut wordiness; and get right to the point and stay there) that can be adopted for writing more clearly and in a simple style. Many authors address issues related with writing in engineering. For example, Budinski (2001) suggests easy-to-follow guidelines, methods and rules that are essential to obtain well-written technical documents. Another interesting source of material is the book by Zobel (1997), more focused on the computer science area.

Requirements writing in a natural language is a research subject. One of the first books to address this subject was written by Kovitz (1999). Another book that deserves to be read, due to the innumerable recommendations that it includes, was written by Alexander and Stevens (2002). Equally relevant is the material described by Robertson and Robertson (2006, Chap. 10), which suggests a generic template for the requirements documents adopted in this book. The article by Knauss and Schneider (2012) proposes an automatic method for aiding organisations (and persons) in learning to write software requirements, based on the requirements documents that they manipulate themselves. The book written by Cohn (2004) discusses various subjects linked with the use of user stories and presents many practical examples about how to adopt them in software development contexts.

Kamsties (2005) illustrates how problematic is the presence of ambiguity in the requirements documents written in a natural language. Yang et al. (2011) present an automatic approach for identifying possible harmful ambiguities, that arise when the text is interpreted distinctly by different readers.

Exercises

Exercise 7.1 (Naveda and Seidman 2006, pp. 29–30) The use of a natural language (for instance, English or Spanish) and intuitive diagrams is used for documenting user requirements. What is the main reason?

1. To eliminate the communication difficulties among the stakeholders.
2. To facilitate understanding.
3. To have no ambiguities.
4. To be precise.

Exercise 7.2 (Naveda and Seidman 2006, pp. 27–28) Which of the following sentences does not constitute a valid system requirement?

1. The product shall be developed using an agile method.
2. The product shall respond to all requests in less than 5 s.
3. The product shall be composed of 12 modules.
4. The product shall use always menu screens to communicate with the users.

Exercise 7.3 Consider the following requirement:

> The system shall be easy to use for trained persons to use.

1. What is the type of this requirement?
2. Is it verifiable? If not, rewrite it so that it is verifiable.

Exercise 7.4 The marketing director has sent you an email message with material for various requirements. Select the relevant parts of the text and rewrite them so that turn into requirements with an appropriate formulation.

> From: Joanne Francis (marketing director)
> To: Fred Wallace (production manager)
> Subject: idea for a game console for cars
> Fred,
> As a result of my meeting with Mr. Shigeru Miyamoto, I think I have identified a new opportunity to extend the products portfolio of our company. We should develop a video game console to be used inside a family car.

The console should turn off when there is no interaction from the passengers for more than 10 min. When the driver is making a phone call, with the help of an hands-free kit, the console should be put in standby mode. The console should weight less than 2 kg, so that it can be easily transported from one car to another one, either by the parents or the children.

The price should be relatively low as a way to be competitive in the market. What is your opinion?

Best regards

   JF

Exercise 7.5 The following are requirements of a controller of a swimming pool:

R1: The input valves should be open when the swimming pool is filling.

R2: The input valves stop when the level of the water reaches 2 m.

R3: The level of the water should increase gradually.

R4: The output valves must be open when the water is higher than 2.1 m.

Which defects this set of requirements present?

(a) ambiguous, (b) solution-dependent, (c) incomplete, (d) not verifiable.

Exercise 7.6 Analyse the following list of user requirements and indicate if they are clear and verifiable. If not, reformulate them so that they possess those properties.

R5: The communication system can only "crash" at most once a month.

R6: The system shall be easy to learn by students with minimal training.

R7: The average delay for the users of a system to pay tolls shall be less than 15 s.

R8: All users shall use the same mobile application.

R9: The maximum delay between the submission of a questionnaire and its confirmation must be half an hour.

R10: The operator shall be able to communicate by mobile phone with the bus driver.

Exercise 7.7 Some of the following requirements present defects that can lead to difficulties in subsequent development phases. Identify those defects and improve the text of each requirement:

R11: The truck driver shall be able to obtain instantaneous answers, if he provides a recognised voice command.

R12: The operator should be able to turn off the system, by unplugging the electrical cable.

R13: The fire extinguishing subsystem must activate when the temperature is over the level of normal operation.

R14: The escalator must stop with no danger to the pedestrians, in the unlikely case of an unexpected fault in the controller system.

Exercise 7.8 Consider a system that permits one to manage the process of submission of computer programming school works (in Ruby) realised by groups of students enrolled in a given course. Each program is submitted by one of the students of the group and is reviewed by the assistants of the professor that coordinates the course. Rewrite, when necessary, the following user requirements in the light of the recommendations addressed in this chapter.

R15: A student must submit in the system his group's program, through a web interface that exhibits soft cores.

R16: The course coordinator assigns an assistant to each submitted Ruby program, in order to revise it.

R17: Typically, the system, itself developed in Ruby, must not assign a program of a group to the assistants that declared conflicts of interest with that group (for example, because one of the members is from his family).

R18: The system shall allow the edition of comments about each program. Those comments will be introduced, as soon as possible, by the respective reviewers. The professor introduces in the system the mark of each program based on the comments. The marks use the 0–10 scale.

Exercise 7.9 Identify for the following sentences the possible interpretations that each one can give rise. Rewrite each one so that it has only one interpretation:

1. Jane has visited Margret's beach house, that left her delighted.
2. The invitation was sent to single aunts and daughters.
3. The banks charge interests to the companies due to delays of their responsibility.
4. The discount applies to children and pensioners.
5. The discount applies to children or pensioners.
6. The parents shall acknowledge the reception of the messages about their child.

Chapter 8
Modelling

Abstract In addition to the use of natural language, the requirements documentation can include formal models for specifying the system in the perspective of those who will build it. This chapter covers topics of modelling in software engineering, with a special focus on the most relevant models for the activities associated with the requirements engineering process. The chapter begins by providing a definition for the 'model' concept. Then, the chapter discusses the ideal characteristics for a model, so that it can meet the goals and purposes defined for its use. The chapter also presents an ontology that allows to accurately relate the various concepts associated with the modelling process. It concludes with a discussion of some structural and behavioural models that are used in the context of software projects, with an emphasis on those that are part of the UML language.

8.1 Definition of Model

Modelling is an essential and inseparable ingredient in all engineering fields as well as a highly creative task. Modelling is the process of identifying adequate concepts and selecting adequate abstractions to construct a model that reflects a given universe of discourse appropriately (Bjørner 2001). Modelling permits the cost-effective use of the model in place of the real-world object or process for some purpose, such as simulation, construction effort estimation, and so on. A model represents the reality for a given purpose, but it is a simplification of reality in the sense that the model cannot represent all its aspects. In fact, if it is to be useful, a model must not represent all aspects of reality. The purpose of the model, which is related to the intention of the modeller (Muller et al. 2012), determines which aspects to represent and which aspects not to. Thus, a model is a form of representing something, not a replication but an intentional selective construction of a new thing meant to stand for something else (Mehrtens 2004). In a nutshell, modelling is related to abstraction, simplification, and formalisation.

Models are simpler, safer and cheaper than reality. A model is not the real world but it is merely a human construct to help a better understanding of a particular perspective of a given system. This allows the world to be dealt with in a simplified

© Springer International Publishing Switzerland 2016

J.M. Fernandes and R.J. Machado, *Requirements in Engineering Projects*,
Lecture Notes in Management and Industrial Engineering,
DOI 10.1007/978-3-319-18597-2_8

manner avoiding the complexity, danger and irreversibility of reality. Models used in engineering are expected to allow one to reason about complex problems and their potential solutions. The main qualities of a model are its abilities to describe, suggest, explain, predict and simulate.

> "If the model can't change reality, reality may change the model."
> *Italo Calvino (1923–1985), novelist*

To be effectively useful, a model of the system under consideration needs to possess to a sufficient degree the following characteristics (Selic 2003):

- **Abstraction**. A model is a reduced description of the system.
- **Understandability**. By removing detail that is irrelevant for a given viewpoint, models, if specified in a form that is intuitive, allows one to more easily understand some of the system properties.
- **Accuracy**. For the properties of interest, a model provides a true-to-life representation of the system.
- **Reasoning**. A model helps with correctly analysing and reasoning about the interesting but non-obvious properties of the system, either through experimentation (e.g., by simulating the model on a computer) or some type of formal analysis.
- **Inexpensiveness**. A model is drastically cheaper to construct and analyse than the system.

Accuracy is not the same as completeness. In fact, models are supposed to be incomplete or approximate in the sense that they omit details that are irrelevant for the given purpose. If the model does not simplify the system under consideration, it becomes useless. Thus, models should be simple, and in accordance to Ockham's Razor, the simplest model that represents the facts is likely to be the most adequate one. Additionally, models are not right or wrong. There is no correct solution that one can check at the end of the modelling process. According to Box and Draper (1987, p. 74), "all models are wrong, but some are useful." This statement reinforces the idea that the legitimacy or adequacy of a model depends on the degree to which it accurately imitates in some way the target system.

The models are considered here as conceptual artefacts. A **model** is an abstraction of the system view and represents a perception or conceptualisation of that system made by the engineer. The inverse operation to abstraction is refinement or concretisation. The model is the result of a first formalisation effort with respect to the system under consideration, and its existence is also at the mental level. The materialisation of the models is made through representations. These questions are handled in detail in Sect. 8.3.

8.2 Model Dimensions

Next, some of the dimensions that can be used for characterising models are presented. These dimensions permit understanding which features a given model can possess. Figure 8.1 shows the three dimensions (form, representativeness, perspective) for classifying models that are considered in this book.

A **physical** (or iconic) model is a reproduction at a reduced scale of a process arising in Nature (Hutter and Jöhnk 2004, p. 394). The word 'physical' here means literally materialised or concrete, since one can indeed touch the model (Illustration 8.1). An iconic model is perceived as imitating the system (looking, sounding, feeling, tasting or smelling like it), being similar with respect to some of its properties (Chandler 2007, pp. 37–38). Iconicity is a straight relation between form and meaning. Typically, a physical model is a smaller representation of the original object (e.g., the Solar system, a neighbourhood, a building, an airplane, an automobile), but sometimes it can be larger if the original object is too small for humans (e.g., an atom, a molecule, a transistor). The interaction with a physical model allows one to obtain information about the properties of the modelled system.

> "The best material model of a cat is another, or preferably the same, cat."
> *Norbert Wiener (1894–1964), mathematician*

Scale models are special physical models in the sense that they seek to preserve the relative proportions (the scale) of the original object. In any case, usually, a physical model is not as accurate or complete as reality. Physical models are not very common in software engineering but occasionally they are employed. One good example is the use of passive storyboarding artefacts made out of paper, Post-it® notes and pencil during the elicitation of requirements (Leffingwell and Widrig 2000, p. 128).

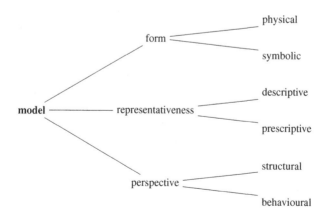

Fig. 8.1 Characteristics of models, according to some dimensions

Symbolic (or mathematical) models use logical and quantitative relationships involving the dimensions of the system (Illustration 8.1). In electrical engineering, models are a set of equations that represent the electrical circuits by applying some laws. An example is Ohm's law ($V = R \times I$), which states that the voltage V (measured in units of Volts) across the terminals of a resistance R (measured in units of Ohms) is proportional to the current I (measured in units of Amperes) through that resistance. Typically, a symbolic model does not resemble the system it represents, but is fundamentally arbitrary or conventional. Arbitrariness, a feature common to many symbolic languages, is the absence of any necessary connection between a linguistic form and its meaning (Trask 1999, p. 9). The relation between a symbol used in the model and the phenomenon of the system it represents must be learnt. For example, a class is represented in UML by a rectangle, but this is a pure convention (Génova et al. 2005). The concept of a class has nothing to do with a rectangle, so a class could be represented by any other symbol. Similarly, the word 'strawberry', composed of 10 letters, has no relation to the fruit itself; it is a pure convention. If this was not the case, there would be no rational explanation to call it 'fresa' in Spanish and 'morango' in Portuguese.

Symbolic models are much more easily changed, if compared with physical models. By manipulating and changing the mathematical relationships, one can see how the model reacts and consequently how the system would react. If the model is relatively simple, then an analytical solution can be obtained by working out a set of mathematical relations. With Ohm's law, if both the voltage and the current are known, the value for the resistance can be calculated. However, for complex systems, the respective mathematical models are also complex which diminishes the possibility of an analytical solution. In these situations, the model can be analysed by **simulation**, i.e., by numerically exercising the inputs of the model to understand how the outputs are affected (Law and Kelton 1991, p. 5).

To create a symbolic model, two things are necessary: (1) a set of signs (or symbols) and (2) a set of rules to operate on those signs. A symbolic model contains a set of representations related to some phenomenon or system. The rules are used to manipulate the symbols and change the model, producing a sequence of representations of the modelled system. If the model is to be useful, then the manipulation rules must be valid not only in the context of the model, but also in the context of the real system.

A **descriptive** model is used to describe or mimic a real-world phenomenon or system. With a descriptive model, one can reason about the properties or the behaviour of the system. An example is a model of the weather that allows meteorologists to forecast it. Since the model is simpler than the reality, reasoning with the model is also cheaper. In almost all natural sciences, like physics, biology, astronomy, and earth sciences, models are descriptive as scientists try to understand how the natural world behaves. In engineering, descriptive models are used, for example, in reverse engineering scenarios when one wants to reason about an existing system without directly affecting it.

A **prescriptive** model is used to define how a yet-to-be-built system is supposed to be. Prescriptive models are adopted in the so-called forward engineering. In software

Illustration 8.1 Symbolic and physical models of the same reality

engineering, models created during the analysis stage describe the problem at hand while design models, typically obtained from those analysis models, represent the system architecture and are used as blueprints for the implementation of the system. So, when prescriptive models are used the target system does not exist yet! It must be engineered. Most of the models used in engineering are prescriptive. Selic (2011) uses the term 'engineering models' to designate models used in all stages (analysis, design, and construction) of the process to develop engineering artefacts.

"We must model the observed world, being it either real or imaginary."
Paulo S. Cougo (1962–), information systems manager

Some models may change from descriptive to prescriptive. For instance, when an architect sketches an old building and then adds some alterations, the model is initially descriptive and later prescriptive (Ludewig 2003). This type of model is called *transient*. Actually, this classification is not a strict property of the model, but rather depends on the relation that exists between the model and the target system.

A **structural model** is focused on the static aspects of a system. These models are used for describing the components or modules that are part of the system, so they serve for conceptualising the system architecture. Class, component, and deployment diagrams, all supported by UML, are examples of diagrams that can be used for representing structural models.

A **behavioural model** emphasises the dynamic, functional, and temporal aspects of the system. This type of models address the behaviour of the system, being thus especially relevant in the analysis phase. Examples of diagrams that can be used to represent behavioural models are finite state machines (Clare 1973; Harel 1987), Petri nets (Murata 1989; Jensen and Kristensen 2009), and data flow diagrams (DFDs) (Stevens et al. 1974).

In addition to these three dimensions, it is also possible to envisage the models according to the type of modelling language used. According to Gajski et al. (1994, p. 19), computer-based systems can be modelled using many different languages, which in general fall into five distinct categories: (1) state-oriented, (2) activity-oriented, (3) structure-oriented, (4) data-oriented, and (5) heterogeneous. These categories reflect the different perspectives that one can have of a system, namely its control sequence, the functionality, or the data structure. The models represented in a given language can be characterised by taking into account the adopted languages. Thus, for instance, if a state-oriented language is used, the model beneath the representation that is being manipulated is also state-oriented.

State-oriented models allow modelling a system as a set of states and a set of transitions. The transitions between states evolve according to some external stimulus. These models are adequate for systems in which the dynamic behaviour is an important perspective to be captured. Finite state machines (FSMs), statecharts and Petri nets are examples of languages adopted to represent state-oriented models.

Activity-oriented models view a system as a set of activities related by data or execution dependencies. These models are well suited to address systems where data are affected by a sequence of transformations at a given rate. DFDs and flowcharts are examples of languages for representing activity-oriented models.

Structure-oriented models allow the representation of the physical modules or components of a system and their interconnections. These models are dedicated to the characterisation of the physical composition of a system, instead of its functionality. Block diagrams, UML deployment and component diagrams are popular examples of languages used to represent structure-oriented models.

Data-oriented models view a system as a collection of data related by some types of attributes. These models dedicate more importance to the organisation of data rather than to the system functionality. UML does not have any kind of diagram that exclusively supports this perspective since it favours object-oriented systems and does not promote the usage of diagrams primarily dedicated to data modelling. Nevertheless, it is possible to argue that UML class diagrams are partially data-oriented models.

When a *data-oriented* model is used, the system is represented as a collection of entities related by attributes, properties, and classes. The models of this type are widely used for developing data-centred systems, where the perspective related to the organisation of the data is the most important aspect to take into account. Entity-relationship diagrams (Chen 1976), that define a system as a set of entities and the respective interconnections, are suitable to represent data-oriented models. Another example of representation for this type of model is the Jackson structure diagram (Davies and Layzell 1993, Chap. 6, pp. 105–113), which organises each datum in terms of its structure, decomposing it hierarchically in subdata.

Heterogeneous models allow for the use of several characteristics from different languages in the same representation. A model is heterogeneous if it incorporates any combination of the characteristics of the four types of models described above. These models are a good solution when relatively complex systems must be modelled. Object process diagrams (OPDs), proposed by Dori (2002), control/data flow graphs (CDFGs), and program state machines (PSMs) are examples of languages for representing heterogeneous models.

Object-oriented models, that can be viewed in an historical perspective as an evolution or extension of the data-oriented models, are *multiple-view* models, since they use simultaneously various models, to address different perspectives (views) of the same system. UML, that appeared as a unifying notation of various development methods, includes a set of diagrams that allow the most relevant aspects of object-oriented systems to be described. Each diagram focuses on a given view of the system and obviously emphasises some aspects and neglects others. The OMT methodology addresses the three following aspects, using for each one a different model: the static structure (object model), the reaction of the objects and the sequence of interactions (dynamic model), and the transformations of data (functional model) (Rumbaugh et al. 1991, p. 149).

8.3 Modelling Ontology

In this section, an ontology that relates and distinguishes some terms commonly used in software modelling is presented. A simplified version of the ontology was initially presented in Machado et al. (2005). An **ontology** is an explicit representation of a shared understanding of the important concepts in some domain of interest. Here, 'concept' is understood as a fundamental category of existence. It is also a mental representation of a class of things (Murphy 2002, p. 1). The ontologies can be represented by conceptual maps, that constitute a simple and well-known form of organisation knowledge (Novak and Cañas 2008). Conceptual maps use concepts, normally represented inside a rectangle, and relations between pairs of concepts, expressed by a line that connects them and by connecting sentences that specify those relations.

The ontology presented here introduces concepts relevant in the engineering domain and it allows to clarify the differences and the relations among concepts like model, specification, description, diagram, language, and notation. This ontology can be seen as a reference model for the "modelling world." The ontology is presented step-by-step throughout the section to allow the reader to comprehend it in a constructive way.

This ontology considers two different and separate perspectives: reverse engineering and forward engineering. In reverse engineering, descriptive models are used to model an existing system, while in forward engineering prescriptive models are employed for modelling a yet-to-be-developed system. This duality assumes that forward engineering is a process of synthesis where the system is developed starting from a model, while reverse engineering is a process of analysis in which the system is seen by means of a model (Génova et al. 2005). This division of the modelling process in two possible alternatives is also considered by Simsion (2007, p. 3), who poses the philosophical question: "Is data modeling better characterized as (a) a descriptive activity, the objective of which is to document some aspect of the real world, or (b) a design activity, the objective of which is to create data structures to meet a set of requirements?" He concludes that the words 'description' and 'design' (in the sense of 'specification' as introduced in this manuscript) are antonyms. The presentation of the ontology is made gradually and initially only the prescriptive perspective is addressed, because it is the most relevant in engineering contexts.

> "A scientist builds in order to learn; an engineer learns in order to build."
> *Frederick P. Brooks Jr. (1931–), computer architect*

8.3.1 System and Model

The two first concepts of the ontology are system and model, already defined in this book. There is a relevant relation between these two concepts, since "a model models a system", as shown in Fig. 8.2. Please notice that whenever a unidirectional relationship is established between two concepts, there exists also the inverse relation. In this case, one could have opted for connecting the system to the model and designate the relation as 'is modelled by'. If one wishes to develop a system that is imagined based on an identified necessity, a prescriptive model is required. The models are considered here as conceptual artefacts, i.e., abstractions of system perspectives that represent the conceptualisation of that system in the engineer mind.

Fig. 8.2 Ontology for the forward engineering perspective: a model is related to a system

The third concept presented in Fig. 8.2 is *idea*. It is a broad notion that intends to encompass the vision, necessities, requirements, and expectations of the stakeholders that induce the development of the system. The idea is necessarily informal and may be spread among the brains of many persons. So, by some form of elicitation, the modeller needs to comprehend the overall idea and devise a model that supports this.

Modelling is the process of obtaining models. It can be considered as a transition from ideas to models, in the prescriptive perspective (or a process that mimics the world in the descriptive one). Modelling is not a totally technical process. Actually, it is a combination of both science and art and includes some sort of creativity. The decisions taken while modelling cannot be totally automated, but tools and guidelines can be supplied to make the process easier and provide a higher likelihood of success.

8.3.2 Specification

To become 'tangible', a model must be expressed by a specification, which is a real representation of the system in a given language. A **specification** (Fig. 8.3) is a formal and declarative representation of a prescriptive model. Specifications are forms of representation that use symbols that stand in for and take the place of something else. A representation is thus a token of the model it represents in the sense that it serves as a visible or tangible manifestation of the model (which is abstract in nature). If an ideal specification could exist, it would posses the following set of qualities: correct, clear, complete, verifiable, consistent, comprehensible by the domain experts, modifiable, traceable, and annotated (Davis 1990, pp. 184–194).

According to Wieringa (2003, p. 39), the outcome of a design decision is documented in a specification which is a representation of the desired properties of the system under development. Due to the use of the expressions 'desired' and 'under

Fig. 8.3 Ontology for the forward engineering perspective: a model is represented by a specification

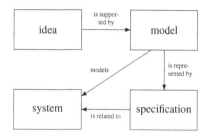

development', this definition necessarily implies forward engineering and consequently prescriptive models. Jackson (2002) refers to 'representation' as the same as model, which is not completely aligned with the proposal made here.

An important concept in the context of modelling is **diagram**, which is a collection of symbols in a two- or three-dimensional space. It is a pictorial, graphical, and visual representation of a logical set of interrelated elements belonging to a model, i.e., it is a partial view of a model. A diagram is thus the expression of a model. For Peirce (1958), a diagram is an iconic model even if there is no rich resemblance between it and the corresponding system, but only an analogy between the relations of the parts of each. A diagram is a representation in which information is indexed by 2D location (Moody 2009).

It is totally acceptable that in some contexts, one uses the terms 'model' and 'specification' interchangeably, e.g., Harel and Politi (1998), Parnas (2010). This relaxation of terms occurs because the specification is indeed a representation of a model. If the language for the specification is adequate for the model that was conceptualised, both the specification and the model convey the same level of information and detail. The specification is the materialised expression of the model.

A model has both tangible and intangible natures. A model is intangible when one considers it from a immaterial point of view, i.e., the meaning point of view. One cannot touch the meaning of a model, though one can touch the symbols that convey the meaning. The symbols used in the description/specification of a model are its tangible expression or manifestation. Thus, a model has also a tangible nature, when it is considered from the material point of view. The linguistic elements used to build up its representation make the model tangible since one can 'touch' lines, boxes, characters, and so on. If one takes 'model' as immaterial, then it is the immaterial meaning of the material diagram or the program code. If one understands 'model' as material (which is also common), then it is the same thing as a diagram or a program, not a different one.

This material/immaterial view of the concept of "message" also existed in the past. According to Gleick (2011, p. 151), a message seemed to be a physical object, i.e., something written on a piece of paper. In the 1840s, when the telegraph was becoming popular, many people had difficulty realising that a given message could be represented in different ways, namely written on a paper or sent by the telegraph through electrical wires according to the Morse code. Gleick tells an anecdote related to a man that brought a "message" into the telegraph office. The operator manipulated the telegraph key and then placed the paper on a hook. The customer complained that the message had not been sent since he could still see it on the hook.

"The medium is the message."

Marshall McLuhan (1911–1980), philosopher

8.3.3 Language

A representation is expressed through a language. According to Aho et al. (2007, p. 118), a **language** is any countable set of strings over a fixed alphabet (set of characters). Since the strings are sequences of symbols, this definition is adequate for text-based languages. A language includes notation and syntactic and semantic rules controlling how to use it.

To be practical, each knowledge representation technique requires a notation (Gašević 2006, p. 5). A **notation** is the set of symbols/signs used for constructing representations. Notation is used to express, in a given representation, the concepts and relations of the domain that apply to the system under consideration. In software, a representation is thus an assemblage of signs (such as words or images) constructed with respect to a notation (concrete syntax) and recorded in a particular medium of communication. The very same model can be represented in many different media. For instance, when one sketches a class, he is actually using the ink of the pencil to draw the sign (a rectangle if UML is adopted) in a given part of the sheet of paper used for that purpose. But, the same class can also be drawn with a computer application. In that case, the class can de "drawn" on the computer screen by using segments of line to obtain the rectangle. Both are specifications (or representations) of the class model that captures a view of the system that is being developed and recorded in different media.

> "A language that doesn't affect the way you think about programming is not worth knowing."
>
> *Alan J. Perlis (1922–1990), computer scientist*

The *syntax* of a language defines its notation, i.e., the available symbols (representing the building blocks, which are the components of the domain) and how one can combine them to express the models. The syntax of a language defines the proper structure of its representations. Syntax is necessary but accidental or arbitrary. *Semantics* defines the meaning of each symbol and how to interpret it with respect to the context of other symbols. In software, the semantics of a programming language establishes the behaviour of the programs.

One usually views languages first syntactically and then semantically. A third aspect, **pragmatics**, studies the meaning derived from context and is seldom considered but, according to Bjørner (2001), it is the most important one followed immediately by semantics. Pragmatics is different from semantics and requires different approaches (Trask 1999, p. 89). Pragmatics refers to practical aspects of how constructs and features of a language may be used to achieve various objectives. For example, the pragmatics of an assignment statement is related to what it is useful for. It may be used to set up a temporary variable for the value of an expression that is required more than once, in order to pass values from one segment of a program to another, to change a data structure, or to set successive values of a variable used in some iterative computation.

Notions of style (or beauty) can be associated with models written in a given language. The notion of style in a programming language expresses the coding practices that yield code easier to read (Monteiro and Fernandes 2006). Whenever there are alternatives in a programming language to achieve a certain result, the one that causes fewer problems for programmers should be considered the one which is in the best style. Many ideas of style appeared due to the advent of superior mechanisms. The idea that GOTO statements are harmful stems from the availability of control structures, like loops (Dijkstra 1968). Fowler (2000) considers the use of the 'switch' statement a code smell (i.e., a symptom in the source code that possibly indicates a deeper problem) due to the availability of polymorphism and dynamic binding. These considerations suggest that the appropriate notion of style for a given language depends on what can be represented with that language.

Therefore, the notions of *notation* and *language* need to be incorporated in the ontology (Fig. 8.4). The language adopts a given notation which must be used by representations. Although it is common to separate languages into two classes (natural and artificial), in this context all languages are considered to be man-made, since no one exists in Nature. Trask (1999, p. 1) asserts that the use of languages is the fundamental issue that differentiates humans from the other animals. When one is manipulating a so-called natural language (i.e., a language which arises in an unpremeditated fashion), like English, Russian, and Arabic, the respective alphabet (Roman, Cyrillic and Arabic) is used as the notation. The letters and symbols of a language can be used to represent all its valid sentences. For the so-called artificial languages, like Esperanto or Java, that notation also needs to be established. For the C programming language, the accepted notation is the set of 128 characters encoded by ASCII, while a Java program may include the characters defined by the Unicode standard.

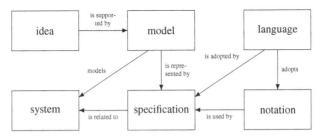

Fig. 8.4 Ontology for the forward engineering perspective: a language and its notation are used in a specification

"The utility of a language as a tool of thought increases with the range of topics it can treat, but decreases with the amount of vocabulary and the complexity of grammatical rules which the user must keep in mind. Economy of notation is therefore important."

Kenneth E. Iverson (1920–2004), computer scientist

The material expression of a model must appeal to some of the human senses, like sight, hearing or touch. Engineering models are usually expressed by either a textual language or a visual language. Textual languages encode information using sequences of characters, while visual languages encode information using spatial arrangements of graphic (and textual) elements (Moody 2009). For a textual language, the symbols that compose a language and the rules, defined in a context-free grammar, for forming valid arrangements of those symbols are its syntax. A visual language is defined in terms of the pictorial signs that constitute its lexical layer. Its abstract syntax is typically described in terms of a meta-model (Guizzardi 2007).

Most software languages are text-based and are used to represent models of software systems. Examples of textual software languages, called programming languages, are C/C++, Java, Python, Ruby, and Perl, and so programs suitable for computer execution can be considered as software models. *Programming languages* are used to describe algorithms in an unambiguous and formal way. Visual languages are non-textual but visible media, including, images, sign languages, maps, and charts. However, in the context of software engineering, a visual language must be understood as a set of valid diagrams (Marriott and Meyer 1998). For example, UML is an industry standard for representing, which means that the information in the models is expressed by graphical symbols (Eriksson et al. 2004, p. 3). More precisely, UML models are expressed through a combination of graphical signs along with textual ones. For example, a UML class diagram represents, in visual form, the structural view of a software system, namely the classes of the system, their names, their attributes, their operations, and the relationships amongst the classes. Other media of expression, rather than text or graphics, are not common. Although one can imagine a software model to be described orally (e.g., being read through the telephone or recorded in an audio file), this is not at all usual since for computer languages one does not define its phonology.

8.3.4 Mental Models

Models encompass many forms and different levels of formalism and abstraction. When an engineer wants to conceptualise a given system, he needs to relate the parts of the system that is to be taken into account. The perception of those parts and their relations, according to a given perspective, define the **mental model** which is a view that has mere existence in the human mind and according to an unstructured and informal representation.

Illustration 8.2 Mental models

The mental process of coming up with a model for a system is not straightforward but rather goes through various iterations permitting it to be gradually constructed. Often, engineers are not conscious of the mental process they follow. Initially, the engineer tends to experiment in a rather random way. Throughout the modelling process, new insights about the model (and consequently about the system) are mentally constructed allowing the engineer to try different alternatives in order to adopt, at some point, a refined solution to pursue. The model, in the engineer's mind, becomes more structured and formal permitting it to be represented in a proper language (Illustration 8.2).

> "What is real? How do you define 'real'? If you're talking about what you can feel, what you can smell, what you can taste and see, then 'real' is simply electrical signals interpreted by your brain."
> *Morpheus, fictional character in "The Matrix" (1999)*

This understanding of models is in accordance with Constructivism; a theory which claims that knowledge is not transmitted, but rather embodied in mental models constructed by each individual (von Glasersfeld 1995). When considering mental models, one really needs to take into account four different things, as noted by Norman (1983): (1) the target system, (2) the conceptual model of that target system, (3) the user's mental model of the target system, and (4) the scientist's conceptualisation of that mental model. According to Norman, a **conceptual model** is created in order to provide an appropriate (in the sense of being accurate, consistent, and complete) representation of the target system. This contrasts with mental models which are kept in the person's mind and usually have many undesired properties like being incomplete, unstable, informal, static, and ill defined. A scientist's conceptualisation of a mental model is a description of a model.

8.3.5 *Model of Computation*

The formalisation of the mental model occurs when it originates a conceptual model that consists in a representation, still conceptual as the name indicates, of the system and according to a particular model of computation. A **model of computation** (MoC) is a collection of rules that govern the semantics of the components and the communication among those components, within a given domain (Lee and Seshia 2011, p. 134). Classical examples of MoCs are dataflow and finite-state machines.

Another important concept (related to this topic) is known as **paradigm** that represents a way of organising knowledge, that is, a way of viewing the world (Floyd 1979). In particular, a programming paradigm is a way of conceptualising what the meaning of performing computation is and how tasks to be executed by a computer are organised and structured (Budd 1997, p. 7). The Sapir-Whorf hypothesis, also known as the principle of linguistic relativity, is relevant here. Basically, this hypothesis asserts that the structure of a language affects, in a large measure, the way in which the humans perceive the world (Trask 1999, p. 46). This hypothesis is widely contested since many believe that it is not viable for natural spoken languages. The Church's conjecture seems also to be not applicable to programming languages. Church's conjecture asserts that from a theoretical point of view, all programming languages are identical, i.e., an algorithm that can be expressed in one programming language can, in theory, be expressed in any other programming language. Both the hypothesis and the conjecture are in a contradiction, as pointed out by Budd (1997, p. 6).

In our opinion, the Sapir-Whorf hypothesis does apply to software languages, in two dimensions. Firstly, modelling languages for software are more general than programming languages, i.e., they are not only used for describing how a given algorithm is to be specified in a formal way so that it can be run on a computer. A modelling language may be used for representing a given perspective of a system. Thus, a particular view, addressed by a modelling language, may not be supported by another language. Secondly, it is well known that polyglot programmers (those who know different programming languages) often approach the same algorithmic problem in radically distinct ways. Iverson (1980), the creator of the APL programming language, also supports this argument by arguing that more powerful notations aid thinking about algorithms. Wexelblat (1980) discusses the effect of the first computer language a person learns with respect to the Sapir-Whorf hypothesis.

One can observe that the definitions of language and MoC are similar. Both mention the components of a domain and the combination of those components to obtain the models or their representations. In fact, a language is the mechanism the software engineer uses to represent models (conceptualised according to a given MoC), using the adopted notation.

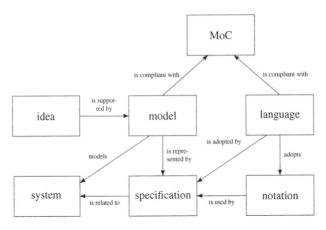

Fig. 8.5 Ontology for the forward engineering perspective: models and languages are compliant with models of computation

"The limits of my language mean the limits of my world."
Ludwig Wittgenstein (1889–1951), philosopher

Figure 8.5 introduces the concept of *MoC* in the ontology, which appears in the third layer of abstraction. MoCs are relevant for both languages and models. The accuracy of a particular modelling approach depends on the capability of the modeller to select the MoC that is best suited to her intention, i.e., the one that semantically best supports the characteristics of the system to be modelled. The modeller has complete freedom to select the MoC in such a way that it clearly conveys the essential features of the model, namely its concepts and relations. Often the modeller does not directly select the MoC, but rather a language that conforms to it. The selected language defines the semantic limits of the system representation at the model level. The characterisation of MoCs is of acute importance due to their impact on the accuracy of the models. If the MoC (and consequently the language) is not adequate, even if one is able to model what is needed, accidental complexity is introduced. If one selects, for example, the MoC for finite-state machines, it will be very difficult, although not impossible, to model a system that includes concurrency issues.

8.3.6 Reverse Engineering Perspective

The explanation of the ontology was made up to this point according to the forward engineering perspective, which is the one that is adopted when the engineer aims to construct a system that does not yet exist. However, the engineer does not want always to build systems from scratch. The **reverse engineering** is the process of analysing a given system to represent it at a higher level of abstraction (Chikofsky and Cross II 1990). In these situations one wishes to analyse or reason about a system that already exists, being thus necessary to obtain a model that describes it. This descriptive perspective, presented in Fig. 8.6, is next analysed, complementing the ontology previously discussed. The differences for this perspective are located in the part that involves the concepts of system, model and description/specification.

The notion of idea is not needed in the descriptive perspective, since one supposes that the system exists (in Nature or after being previously created by humans) and that the engineer can observe it to devise a model. One supposes that the things of the universe, not created by man, simply exist. *Modelling* is the process of obtaining models and can be considered as a transition from ideas into models in the forward engineering perspective or a process that imitates the world in the reverse engineering perspective.

Another difference is that in the reverse engineering approach, the model should be represented by a *description* (and not a specification) to be materialised. A **description** is a representation of a descriptive model. The descriptions and specifications can be generically referred to as representations.

The last difference is that the relation between system and model has the opposite direction and uses a sentence in the passive voice ("is modelled by" instead of "models"). Even though this change was not strictly necessary, it permits highlighting the existence of the system before the model. In the forward engineering perspective, the relation is inverted to make it evident that the model exists before the system.

According to the terminology used here, modelling corresponds to the activity of selecting a MoC in order to formalise, at the conceptual level, a given target system, while specification and description are related to the adoption of a language to represent, in a tangible form, the model of that system. Ideally, in science and

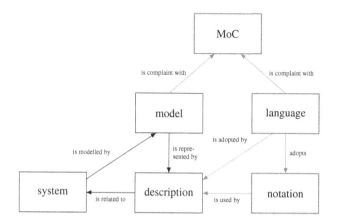

Fig. 8.6 Ontology for the reverse engineering perspective

engineering the MoCs and the languages should be consciously selected based on the characteristics of the target system and the modelling purpose (or the intention of the modeller).

Hence, engineering models are here classified as mental, in the sense that they only exist in the mind of the developers. The very same model can be represented in many different ways depending on the adopted languages. These different representations are called specifications or descriptions if they refer respectively to prescriptive or descriptive models.

Obtaining a specification or a description, that adequately represents the system, depends on both the characteristics of the selected MoC for the modelling activity and the MoC of the chosen representation language. Thus, to avoid semantic mismatches and incompatibilities, the two MoCs must be compatible (as close as possible). Ideally, the MoC of the language should be the same as the one used in the system modelling activity. In this context, it becomes clear that the characterisation of MoCs is a fundamental issue for accomplishing both the modelling and specification/description activities.

8.3.7 Analogies

In order to facilitate the grasping of the discussion made in this chapter and allow for its contents to be comprehended by readers not familiar with software engineering, this subsection provides some contexts where analogies can be drawn with other fields. The first analogy drawn is related to numbers. Numbers are like conceptual models in the sense that humans are able to master them in their brain. The concept (model) of the natural number four (object of interest) is somewhat independent of the form (language) with which it is written (represented). The concept of the number 4 can be represented in diverse ways: '4' (Hindu-Arabic numerals), 'IV' (Roman numerals), '100' (binary system), '||||' (Greek numerals), 'four' (English language), 'quatro' (Portuguese language). The choice of the language is relevant for an adequate representation, since, for example, zero or fractional numbers cannot be represented by Roman numerals. For example, complex numbers constitute a different (i.e., larger) domain when compared with natural numbers, and consequently require a different (i.e., more powerful) language.

The second analogy is related to music. Rowell (1983) defines music as referring to sounds, a piece of paper, an abstract formal concept, a collective behaviour of society, or a single coordinated pattern of neuro-chemical impulses in the brain. This definition is very broad and covers many aspects, forms, and both processes and products. The creator of a music piece needs to conceptualise it in his mind. A piece of music can be represented on a sheet music (also designated score) according to a musical notation that uses written symbols. But as Alperson (1994) puts it: aren't scores silent? If one is able to play an instrument or sing, sound waves are sent through the air to a human ear and the music becomes tangible (i.e., aurally perceived). The existence of a piece of music does not require it to be played. The conceptual nature of music can be exemplified by the fact that Ludwig van Beethoven was almost completely deaf when he composed his ninth symphony, considered as a supreme masterpiece (Cook 1993). So, Beethoven's mind was able to conceptualise the symphony and represent it on a sheet music, even if he never heard it.

> "If I were not a physicist, I would probably be a musician. I often think in music. I live my daydreams in music. I see my life in terms of music."
> *Albert Einstein (1879–1955), physicist*

Another analogy can be established with respect to money. Money is a social creation, an abstraction that serves as a unit of account to assign value to goods and services. Modern societies use fiat money, whose value derives from law or governmental regulation. The term 'cash' refers to money in its tangible form; the monetary value of coins and banknotes depends almost entirely on a shared understanding (Baumeister 2005, p. 61). In fact, cash is a form of representative money, since it has no intrinsic value, i.e., the face value of a banknote is (much) greater than the value of its material substance. Contrarily, commodity money consists of objects that have value in themselves, as well as value as a currency. Examples of commodities used as tokens of exchange include gold, silver, alcohol, fur, or salt. A silver coin is not a symbol of money, but an incarnation of it. Nowadays, money is becoming even more abstract, more virtual, and more digital. These characteristics are well demonstrated by Bitcoin, a recent digital currency (Peck 2012). Advances in computer technology, telecommunications, smartphones, and cryptography provide the opportunity for a cashless (or less-cash) society, where the necessity to physically manipulate bills and coins is decreasing. However, as pointed out by Smithin (2000), this is just a change of form (i.e., representation) rather than substance.

8.4 Models for Requirements

Modelling, done during the analysis phase, aims to specify the requirements of the systems. Nowadays, most software models are, in industrial contexts, represented in UML. This language presents many types of diagrams that support the description and specification of different types of models. However, as discussed by Erickson and Siau (2007), most development teams use in practice only a subset of those diagrams and from those also resort just to part of their available constructors. This situation is easily framed in the Pareto law (or 80/20 rule): 80 % of the modellers use just 20 % of the UML constructors.

Table 8.1 identifies the six models that we consider as essential for documenting the various aspects that are important when modelling software systems. The table shows also, in a succinct way, the purpose of each one, i.e., the perspective of the system that each one highlights. In the following subsections, there is a short description of those six models, illustrating the most relevant modelling concepts that each one

Table 8.1 Purposes of the main models used for developing software-intensive systems

Model	Purpose
Domain	Describe the vocabulary, concepts of the domain and characteristics of the systems that can be developed for the considered domain
Use case	Describes the proposed functionalities of a given system
Interaction	Show how the various objects or entities collaborate, emphasising the flow of control and data among them
Class	Present a set of concepts, types and classes and the respective relations
State	Specify the behaviour of an entity or indicate the various states (or modes) through which it transits throughout its life
Activity	Show the control flow among the activities of a process

covers. For such, the UML diagrams that support the representation of those types of models are discussed. Although UML can be used in various contexts, it was specifically conceived for object-oriented systems, hence, in parts of the discussions made next, that reality is sometimes reflected in the text.

8.4.1 Domain Models

As mentioned in Sect. 2.2.3, many software systems are focused on a given domain. In those cases, it is relevant to use domain models for capturing the common elements of the systems to be developed. This utilisation should happen, either when the producer approaches for the first time that domain, or when it works continuously in that domain. In the former case, the model aids the team to rapidly gain knowledge about the domain; in the latter case, it permits one to construct a repository of information that serves as reference for all the system to develop.

A *domain model* constitutes a description of the common properties and variables of the domain related to the system that is being developed. The domain model expresses enduring truths about the universe that is relevant to the system at hand (Fairbanks 2010, p. 115). That description must include (1) a definition of the scope of that domain, providing examples of systems or generic rules of inclusion, (2) the vocabulary of the domain (i.e., the glossary with the principal terms), and (3) a model of concepts that identifies and relates the concepts of that domain. This model represents the things (entities or events) that exist in that domain, that is, it is a conceptual reference of the problem domain. An important advantage of a domain model is to permit describing and restricting the scope of the problem domain. Thus,

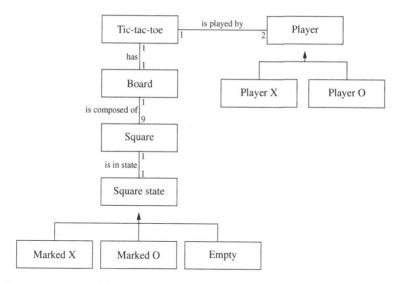

Fig. 8.7 A domain model for tic-tac-toe

the domain model can be used for verifying and validating the understanding that the stakeholders have about the problem domain.

Figure 8.7 shows the domain model, represented by a class diagram, for the tic-tac-toe game.

8.4.2 Use Case Models

The utilisation of use case models serves essentially two purposes: (1) defining the frontier of the system with the environment, and (2) specifying the functionalities that the system makes available to its users. A *use case diagram* resorts to, as basic elements, *use cases* (represented by ellipses) and *actors* (represented by stylised human figures), indicating, respectively, which functionalities that system offers and what exists outside it. Figure 8.8 shows an example of a use case diagram for a lift. The usage of verbs is recommended to characterise the use cases, thus enhancing their functional nature.

A **use case** describes a set of actions, executed by the actors and by the system, so that a valuable result is produced to the users (Booch et al. 1999, p. 468). By valuable, one understands here as solving a given problem or satisfying a given requirement. In other words, a use case defines a series of interactions between the system and the actors (humans or external systems) that allows a given result or objective to be achieved. That result is associated with the satisfaction of the necessity that is in the origin of the system development. Usually, use cases are supplemented

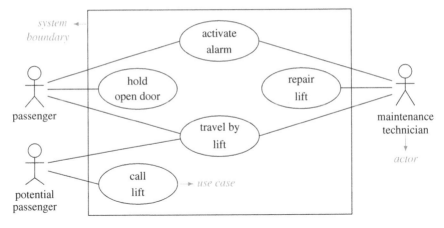

Fig. 8.8 A use case diagram

with scenarios, that allow the sequences of actions performed to be described (see Sect. 5.3.3).

The use cases of a system constitute a functional decomposition of the behaviour of that system, without imposing it any internal structure. A use case diagram shows the functionalities of the system, without restricting the sequentialisation of those functionalities. One can infer that use case 'travel by lift' occurs after use case 'call lift', but that is only a deduction. The use case constitute an important model, in the context of the software development, since it allows to capture, from the stakeholders, the user requirements. The use case are generally initiated by an actor, although sometimes it is useful to be the system itself to initiate them (for example, periodic calculation of the interests to be credited in bank accounts). UML gives great importance to use case diagrams, since it is based on them that, according to its proponents, one can plan and support all the development of the system under consideration.

A use case must be detailed through textual descriptions. There are several proposals on how to describe a use case. Here, one suggests, regardless of the format, to consider the following fields: (1) identifier, (2) designation, (3) short description, (4) pre-conditions, (5) actors, (6) main scenario, (7) alternative scenarios, (8) comments, (9) date, and (10) version.

An **actor** represents a role that the users can have with respect to a given system, when interacting with it. The symbol for the actor has the aspect of a very stylised human figure, but that does not imply that the actors are necessarily human. Actors cam also represent other systems or entities (for instance, a bank) that externally interact with the system at hand. Various persons can be represented by the same actor, as can be observed by the fact that the lift was built to serve various passengers and not just one. An actor is an abstract representation of a type of person that interacts with the system.

"Use cases have proven particularly valuable as part of the requirements activities of the software process."

Ivar Jacobson (1939–), software engineer

Use case are executed by the actors and the same actor can be involved in various use cases. It is also possible that a given use case implicates various actors. The same person can perform various roles with respect to the same system; those roles give origin to distinct actors in the respective model. In the lift example, the maintenance technician can also, in some situations, be a passenger. In a more complex example, the same person can be a manager, shareholder and customer of a bank.

The identification of the system actors eases the definition of the functionalities, made through the identification of the use cases. A use case consists in a particular way of using the system, representing part of its total functionality. Each use case constitutes a complete set of events, initiated by an actor, and clarifies the interaction that one can observe between that actor, the system and, possibly, other actors. The set of all use cases permits us to specify all the distinct forms of interacting with the system.

When one defines the actors and the use cases that are part of the system, one is delimiting the system, i.e., is defining the scope of the system that is to be developed, which is important at the beginning of the development process. Many software methodologies, for instance (Dennis et al. 2009, p. 177; Shelly and Rosenblatt 2011, p. 208), suggest as one of the first analysis tasks, the creation of a *context diagram*, whose first utility is to clearly delimit the frontier of the system and show which entities interact with it. A context diagram defines unequivocally the environment that surrounds the system under development and, when using use cases the creation of the context diagram is being realised to a large extent. Figure 8.9 depicts the context diagram of the lift.

A related term to use cases is scenario. In UML, a *scenario* refers to a give pathway inside a use case, given by a specific combination of conditions. A scenario is an instance of a use case, in the same way as an object represents an instance of a given

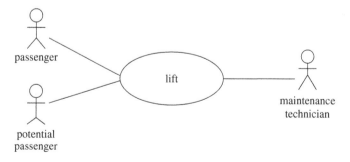

Fig. 8.9 A use case diagram to represent the context of a lift

class. The scenarios can be modelled in UML, through sequence or collaboration diagrams (see Sect. 8.4.4). This means that, for specifying a use case, it is necessary to consider various scenarios. As Fig. 8.8 illustrates for the lift, the "call lift" use case represents the call of the lift by a potential passenger. This use case can be associated with various scenarios, like the following ones:

- the lift is stopped at the same floor as the passenger;
- the lift is stopped, but at a different floor than the one where the passenger is;
- the lift is moving up and will pass the floor where the passenger is waiting to go down;
- the lift is moving down and will pass the floor where the passenger is waiting to go down;
- the lift is moving but will not pass the floor where the passenger is;
- the alarm was actuated by another passenger who is inside the lift.

Each use case is related to one or more requirements. The most complex use cases are associated with many requirements, while the simplest ones are related to a reduced number of requirements. The requirements can be the elements that are used for defining the steps of the use case scenarios.

8.4.3 Class Models

For object-oriented systems, *class models* are necessary to indicate the existing classes and their relations. These models are, in some of their variants, always contemplated by all object-oriented software development methods, since the concept of class is fundamental in this paradigm. A class can be represented as Fig. 8.10 illustrates. The class is divided into three parts: the top part is used to indicate the class name, the central part indicates the class attributes, and the bottom part lists the class operations. The name is mandatory, but the other parts can be omitted.

A class diagram aims to highlight the static structure of the concepts, types, and classes. The concepts show how the users see the solution domain, regardless of the form in which those concepts are actually implemented. An example of this kind of diagram is shown in Fig. 8.11. Class diagrams allow us also to describe the types of objects (the classes) that the system may contemplate and the forms of

Fig. 8.10 A class for a student

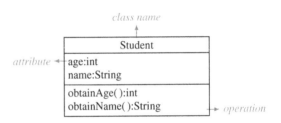

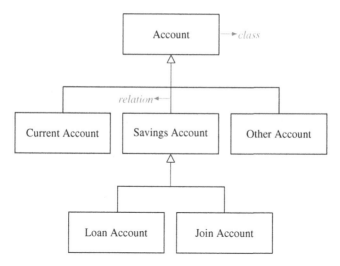

Fig. 8.11 An example of a class diagram, involving different types of bank accounts

interrelationship among them. In this example, the attributes and operations of each class are omitted, since that indication was deemed unnecessary.

The message is the fundamental unit of communication among objects. If two objects communicate, then there is a connection or link between them. There are in UML basically four types of relations between objects, that can be shown between the classes in the respective diagrams:

- **Association** represents a relation between objects that manifests itself at execution time through the exchange of messages (for example, a student enrolls in an university or a company employs several workers). When a class of objects depends on the services of another class, they should be connected by an association. The associations are represented in UML by lines and by default they are bidirectional.
- **Aggregation** represents the so-called part-of relationship, that occurs when an object contains another, either physically or logically. The included class is called a *component* or a *part* and the wider class (the one that includes the components) is called an *aggregate* or a *composite*. In UML, an aggregation is represented by a line with an unfilled diamond near the aggregate. It is possible that the components can be shared among various aggregates, if deemed adequate.
- **Composition** constitutes a more restricted form of aggregation, in which the components are shown by graphical inclusion in the aggregate; alternatively, a filled diamond can be used, but the first representation is preferable. The components of an aggregate by composition cannot be shared with other aggregates. The aggregate is responsible for creating and destroying its components.
- **Generalisation** is used when a class is a specialisation of another class. The subclass inherits all the characteristics of the superclass and can add new attributes

or operations. In UML, this relation between classes is represented by an arrow that begins at the subclass and ends at the superclass.

> "The first step in wisdom is to know the things themselves; this notion consists in having a true idea of the objects; objects are distinguished and known by classifying them methodically and giving them appropriate names. Therefore, classification and name-giving will be the foundation of our science."
>
> *Carolus Linnaeus (1707–1778), botanist*

Figure 8.12 shows the relations among classes previously described. Each relation has two *roles*, representing each one the form in which each object participates in the relation and providing a possible reading of that relation. The inclusion of the roles is optional, but it is recommended when it clarifies the relation between the objects. The use of the roles becomes necessary for relations between objects of the same class or when two objects have more than one relationship. Generally, it is not necessary to identify the two roles of the same relation, since one is the conjugated or opposite of the other. In the example, the Computer is the master of the Controller, while the Controller is the slave of the Computer.

Alternatively, a *navigation arrow*, represented by a small solid triangle, can be placed to name the binary association and to show the order of the ends of the association. The arrow points along the line in the direction of the last end in the order of the association ends. This notation also indicates that the association is to be

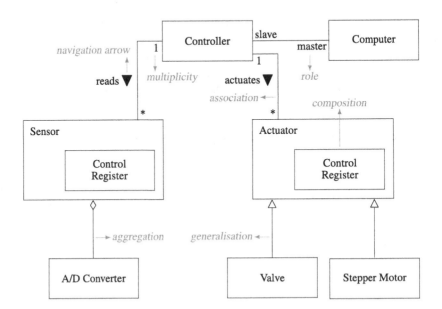

Fig. 8.12 The various possible relations between classes, according to UML

read from the first end to the last end. So, in Fig. 8.12, the relation between Controller and Actuator is read "Controller actuates Actuator".

The numbers or symbols at the end of the relations, designated as *multiplicity*, indicate the number of objects of the respective class that participate in the relation. The multiplicity indicates the upper and lower limits for the number of objects that participate in the relation.

8.4.4 Sequence Models

In same cases, it is necessary to model the dynamic aspects related to the exchange of messages between objects. Those models can be represented by *interaction diagrams*. As shown later, state models, associated normally to classes, do not fit this purpose, since they just describe the internal state changes of the instances of a given class and not what happens among a set of objects (usually from different classes).

An interaction model can be used for representing an instance of a use case. They describe how a group of objects communicate amongst them. Normally, an interaction model captures the behaviour of a scenario of a given use case, showing the objects and the messages that are exchanged among them. In UML (version 2.2), there are four different types of diagrams that allow interaction models to be represented. Since, in their essence, all those types address the same modelling purposes and are equivalent in terms of what can be modelled, this book presents only sequence diagrams.

Sequence models can be used to describe the behaviour of the system. They can also be used, during the system testing, for comparing the real behaviour of the system (prototype or executable model) with the one that was specified. A *sequence diagram* shows a sequence of messages exchanged among objects. The principal elements that can be found in this type of diagram are shown in the example of Fig. 8.13.

A vertical line, called *lifeline*, represents an *object*, being the respective name indicated above or below that line. A horizontal arrow is a *message*. Each one has its origin in an object (responsible for its creation) and ends in another object (to which the message is addressed). The indication of the *name of the message* is done near (usually above) the arrow.

Temporally speaking, the diagram is read from top to bottom, which means that in Fig. 8.13, for example, message "request to go up" is sent before message "request to go down". It is important to note that the temporal axis is not associated with any scale, showing just the order (before or after) between the events. On the left side of the diagram, one can include some *textual annotations*, with the purpose, among others, to identify initial conditions, actions, and activities not evident just by reading the diagram.

Most sequence diagrams contain only the elements indicated previously: objects, messages, periods of inactivity, and textual annotations. The sequence diagrams contain three additional mechanisms that are useful for modelling some types of

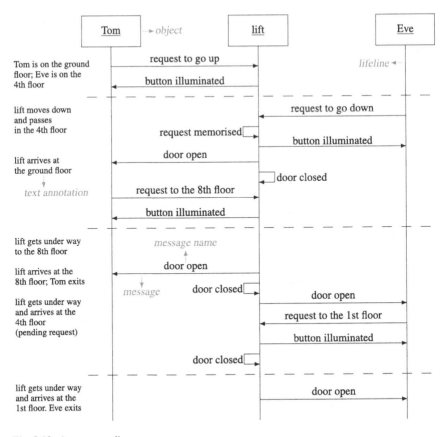

Fig. 8.13 A sequence diagram

systems: event identifiers, timing marks, and state marks. These mechanisms are explained with the example that Fig. 8.14 illustrates.

Near a message, one can include an *event identifier* that indicates which event is responsible for sending that message. For example, the figure shows that the delivery of message msg1 is due to the occurrence of event ev1.

For real-time systems, in which there are time restrictions to consider, it becomes indispensable to use *timing marks*. There are two different forms of indicating them. The first one consists in using a bar with a time limit, defining thus the time that it takes to go from the events that are in the extremes of the bar. In the figure, this form is used to record that the time between events ev4 and ev5 is 2 s. The second form consists in indicating between the brackets relational expressions that specify time restrictions. This form is used in the figure to indicate that the time between events ev1 and ev2 cannot exceed 15 s.

It is also possible to add *state marks* to the sequence diagrams. State marks, represented by rounded-corner rectangles (the same symbol for states in state machines),

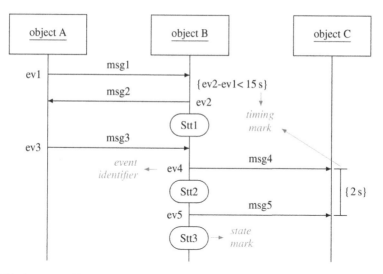

Fig. 8.14 A more sophisticated sequence diagram

are put in the lifeline of a given object, allowing one to identify the various states in which that object can be found. It is a form of relating, more easily, the sequence diagrams with the diagrams state. For object B in that figure, three distinct states are identified: Stt1, Stt2, and Stt3.

8.4.5 State Models

Class models do not allow the dynamic behaviour of the instances of the classes to be determined. Thus, for some more complex classes, the use of a notation that permits us to address state-oriented behavioural modelling is crucial. Historically, the use of state diagrams has become popularised in the hardware domain, but its modelling capacity has proved to be useful in diverse computing areas (for example, software, compilers, communications, operating systems, simulation, multimedia, human-machine interfaces).

State diagrams can be used for defining the (dynamic, temporal) behaviour of a class (i.e., its instances). They allow us to detail all the states in which each one of those objects can be found and the transitions between states triggered by conditions to which those objects are sensitive. In a conventional state diagram, one and only one state is active in each instant. A system state is represented by a set of variables, whose values contain all the information necessary about the past of the system and that simultaneously restrict the future system behaviour. A state represents a period of time during which the system exhibits a specific type of behaviour. A **state** is an ontological condition that persists for a significant period of time that is distinguishable and disjoint from other similar conditions (Douglass 2004, p. 402). A state differs from other states in the events it accepts, the transitions it takes as a result of the accepted events, or the actions it performs. A transition is an answer to an event that causes a state change.

> "A man provided with paper, pencil, and rubber, and subject to strict discipline, is in effect a universal machine."
>
> *Alan M. Turing (1912–1954), mathematician*

State machines are used when a transition between states occurs, mainly as an answer to significant events. State machines extend the most conventional diagrams in three axes related to hierarchy, concurrency, and communication (Harel 1987). Despite the fact that these extensions permit one to obtain simpler, more compact, and more readable models, UML state machines are mathematically equivalent to conventional state machines (either Moore or Mealy machines).

Figure 8.15 shows an example of a state machine, which is composed of two main elements, the *states* and the *transitions*, represented respectively by rounded-corner rectangles and arrows. The diagram shows also superstates Stt1 and Stt3. A *superstate* represents a state that contains other states inside its contour, constituting the mechanism to hierarchically structure a state machine. This abstraction mechanism permits a state machine to be constructed and seen at the desired level of detail.

Transitions connect any two states, regardless of the different nesting levels in which they may be. For example, t6 causes the transition from state Stt31 to state Stt32. In case of transitions t1 and t5, due to the fact that they terminate in superstates (Stt1 and Stt3, respectively), it is necessary to apply some rule for determining which substate is activated. An alternative is to identify one of the states as initial, using a black circle and an arrow: state Stt32 is the initial state of superstate Stt3. Another solution is to use the *history connector*, as is done in the superstate Stt1 with respect to substate Stt12. In this case, transition t1 determines that the substate to be activated must be the last one in which Stt1 was active before exiting it in the last time (in the example, either Stt11 or Stt12). This mechanism constitutes a way of memorising

Fig. 8.15 A state machine

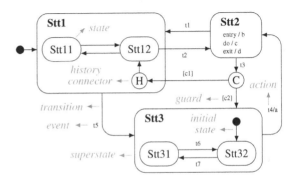

the last active substate that needs to be resumed as soon as the respective superstate is re-entered. If there is no history for the superstate (because it has not been active or has previously reached the final pseudo-state), the history connector points to the initial state by default.

The labels on the transitions can be composed of three elements, all of them optional, according to the following structure: *event [guard] / action*. The name of the event that triggers the firing of the transition is "event", the logical expression that has to be evaluated as true for that transition to fire is "guard", and the list of the operations that are executed as a result of the transition being fired is "action". The actions include, in addition to simple commands (for example, the assignment of a value to a variable: `cont = 5`) or calls to operations, lists with the generated events, which are propagated to all the system objects sensitive to them.

In UML, an *event* has a wider scope than the one that usually is associated with it and represents an occurrence whose consequences are important for the system. There are four different types of events in UML:

- **call event**: Represents a synchronous and explicit request from an object to another one, waiting the former for an answer from the latter. It is a form of invoking from a method from an object, seen as an implementation of an operation.
- **change event**: Expresses the change in the value of a given logical expression (when it changes from false to true). This type of events imply a continuous test, which in practice can be relaxed with checks in specific moments, as long as one makes sure that no occurrence is lost.
- **signal event**: Describes the reception of an (identified and explicit) asynchronous signal for communication among objects.
- **time event**: Expresses the arrival of the absolute time or the passage of a relative time. In real implementations, the time events do not come from the universe, but rather from some clock object (component) available inside or outside the system.

Whenever a transition ends up at a superstate contour, one (and only one) of its substates has to be activated. Additionally, a transition that starts at a superstate contour corresponds to a graphical simplification that represents the application of that transition to all its substates. This last mechanism, associated with the history connector, is extremely useful, for instance, to model exceptional conditions that deserve to be handled in any situation, with the system returning to the previous state, after the provided handling. The state machine shown in Fig. 8.16a intends to illustrate the graphical simplification that results from the utilisation of these two modelling mechanisms, in comparison with the state machine in Fig. 8.16b, where those mechanisms are not used.

Taking the state machine in Fig. 8.15 again, one notices that transition t3 ends at a conditional connector (represented by a circle with a C). It is a transition whose next state depends on the evaluation of logical conditions, that is, the guards associated with the transitions. If the guard [c1] is evaluated as true, then the next state is Stt11 or Stt12. Due to the history connector, it will be the last of these states that were active in the last time that superstate Stt1 was exited. If the guard [c2] is true, then the next state is Stt32. In order for the state machine to be deterministic, it is necessary to make

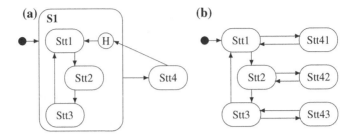

Fig. 8.16 Equivalent state machines for handling exception conditions

sure that the conditions are complete (i.e., that all the possibilities are considered) and mutually exclusive (i.e., that two or more guards are never simultaneously enabled). The use of the [else] guard can be considered for ensuring these two conditions.

Consider now Fig. 8.17 to show a few additional mechanisms of state machines. The behaviour begins at the black point (on the left side of the figure), which means that the first state activated is "verifying". That transition presupposes the execution of the "search first item" action. The highest level initial transition of a state machine, which is triggered when the respective object is created, cannot have any associated event. It is only allowed to specify an action that is executed as soon as the object has just been created. When this action is complete, the system state becomes "verifying", which is associated with the "verify item" activity.

In UML, action and activity are distinguished concepts. Both represent processes that must be executed and are typically operations provided by the object under consideration or the objects that communicate with it. An *action* can be linked with transitions and is considered as an instantaneous process, that happens very rapidly and that cannot be interrupted. An *activity* is associated with states and normally is a

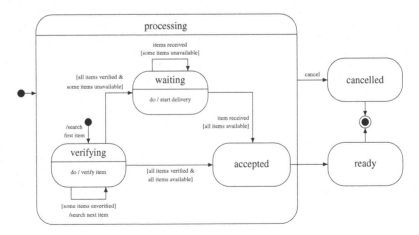

Fig. 8.17 Another state machine

longer operation, being possible to interrupt it. States can be associated with *entry* or *exit actions*, as well as activities. The entry actions are operations that are executed when the respective state is entered, and the exit actions are executed when that state is left. The activities, indicated by the word *do*, are executed while the state is active. In the state machine shown in Fig. 8.15, state Stt2 has these three types of operations.

State machines can be constructed incrementally. To exemplify this possibility, consider the state machine in Fig. 8.18 that represents a refinement of the one presented in Fig. 8.17. The labels in the transitions are omitted for the sake of simplicity. The behaviour starts in the block point (on the left side of the figure), which means that the first active state is superstate "processing". The dotted line separates two orthogonal regions, that represent two independent behaviours and that introduce concurrency in the model.

When the machine is in state "processing", it is mandatory to be exactly in two substates, one in each one of its regions (for example, "verifying" and "authorising"). Although this modelling does not necessarily imply concurrency or parallelism among threads of execution, such mechanism is one of the possible solutions for implementing those cases. The orthogonal regions can communicate among them, through events or guards dependent on common variables, in order to synchronise their behaviours.

In addition to this modelling perspective more oriented towards behaviour, state machines can also be used to model the states or modes of a given relevant entity. Figure 8.19 shows a state model for the civil status of a person (single, married, divorced, or widower).

> "The only really happy folk are married women and single men."
> *Henry Louis Mencken (1880–1956), journalist*

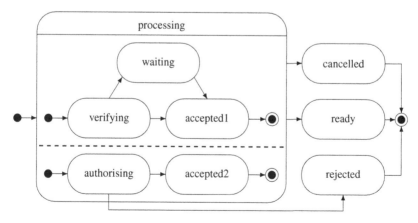

Fig. 8.18 A state machine with concurrency

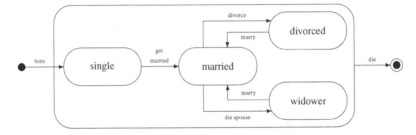

Fig. 8.19 State model for the civil status of a Portuguese citizen.

8.4.6 Activity Models

Activity models are useful to relate the control flow among the activities of a given business process. These models address behavioural aspects of the systems or entities under consideration. These models are appropriate when the behaviour change occurs, mainly due to the end of the action/activity executed and not to the occurrence of events, as is the case with state models.

Figure 8.20 presents an example of an activity diagram that models the process of collecting products in a warehouse. The process starts with the activity of picking products (based on a list). As soon as this activity is finished, one can then check if

Fig. 8.20 Activity model for the process of collecting products (initial version)

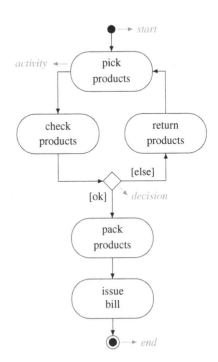

the collected products match exactly with the ones that are in the list. This activity assumes a decision: if the products match, then the products can be packed and later a bill is issued; otherwise, the products that are not part of the list are returned. Next, one must repeat the picking activity to obtain the missing products. The pick-check-return cycle repeats the necessary number of times, until the products are all correctly picked.

This model can be enriched. Figure 8.21 illustrates the same process, but now with the identification of the different departments responsible for each activity. Each swim lane (or partition) identifies the entities responsible for the execution of the respective activities. The activities of packing and billing can be executed in parallel, since there are no dependencies between them and they are executed by distinct entities. This small example shows that modelling can be done in a progressive way, adding details that turn the models richer. The presence of the swim lanes is an indicator that it is convenient, but not compulsory, to draw the activity diagrams from top to bottom, namely when a given process is being modelled. This recommendation permits also to read and interpret the diagrams in a more uniform way.

The choice between a state model and an activity model is not always easy to take. The first recommendation is that any of those models should just be created for entities that have a significantly elaborated behaviour. For example, it does not make sense to create a state model for an entity that commutes just between two states: active and inactive. The choice of the (state or activity) model depends on the nature of the

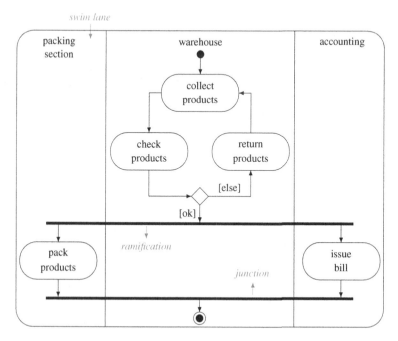

Fig. 8.21 Activity model for the process of collecting products (improved version)

entity. A state (or mode) does not necessarily imply that some activity is in execution. For example, an equipment in the state 'off' does not perform any activity. State models are normally used for entities that go through different states (for instance, the civil state of a person or the state of a library book). The transition of states is made based on events that occur in the context of the entity under consideration (for example, one changes from married to widowed, when the spouse dies). Activity models are relevant for illustrating the relations among the various tasks or activities that constitute a given process. A transition from an activity to another one frequently occurs only when the first activity ends (for example, in the process of collecting products in the warehouse, they can only be packed after being properly checked).

8.5 Summary

Using models is essential in all engineering branches. A model represents in a simplified way the reality for a given purpose, emphasising some elements and ignoring others. The models to be effectively useful, must possess to a sufficient level the following characteristics: abstract, comprehensible, accurate, predictable, and inexpensive. The models can be characterised according to three dimensions: form (symbolic, physical), representativeness (prescriptive, descriptive), and perspective (behavioural, structural).

The chapter presents an ontology that introduces relevant concepts related to modelling and permits one to distinguish and associate concepts like model, specification, description, diagram, language, and notation. This ontology is a reference for the 'engineering modelling' universe. It considers a dichotomous division of the world, through two different perspectives: reverse and forward engineering. This chapter also presents some UML models for documenting the requirements of a system. For requirements engineering, domain, use case, interaction, class, state, and activity models are deemed crucial.

Further Reading

The ontology presented here contrasts with the terminology presented by many authors. For example, Kühne (2005) defines a model as an artefact formulated in a modelling language, describing a system through the help of various types of diagrams. Jackson (2006) uses the term 'model' for a description of a software abstraction, but the notion seems to include both the reverse and forward engineering perspectives (i.e., descriptions and specifications). Seidewitz (2003) classifies models into two categories: descriptions and specifications. A model is either a description of something that exists or a specification of something that must exist.

This chapter presents in a succinct way various UML mechanisms and diagrams. The language is too large with so many mechanisms, so the interested reader should look for user guides or manuals, like the one written by the creators of the language (Booch et al. 1999). The books (Fowler 2004; Eriksson et al. 2004; Stevens and Pooley 2006; Seidl et al. 2014) also cover UML, a language in a permanent update.

Use cases can be further explored, by reading the material presented by Jacobson et al. (1992) and Schneider and Winters (1998). Cockburn (2000) is worth consulting, since it is succinct, provides recommendations on how to structure use cases, and advises when to stop writing them. Jørgensen et al. (2009) present executable use cases, which constitute a model-based approach to requirements engineering and link user-level requirements and more technical software specifications. More complete information about state machines, namely some modelling mechanisms not presented here, can be found in (Harel 1987; Harel and Gery 1997; Douglass 2004; Rumbaugh et al. 1999).

Exercises

Exercise 8.1 Construct a domain model for:

- the draughts (or checkers) game;
- chess.

Exercise 8.2 Construct an activity diagram that models the set of tasks, performed by the (male and female) students of a college, since when they take a shower until they dress. The minimum set of tasks to include in the diagram is: dress up shirt, wear pants/skirt, wear underwear, put shoes on, brush teeth, take shower, and shave. The following set of rules must be respected:

- the hygiene tasks, except for the shower, can be executed simultaneously with the dressing tasks;
- obviously, only boys should shave themselves;
- girls may use either pants or skirts;
- when a student is dressing up, he can change the socks if he feels that they do not match with the shirt.

Exercise 8.3 Consider that a book in a library can be in one of the following states, throughout its existence: received, ready to be catalogued, available for consulting (at the library), available for requisition, blocked, borrowed, missing.

- Model the states of the entity 'book' with an adequate diagram.
- Add to the diagram the 'reserved' state. Assume that a book can have various persons in a waiting list.
- Add to the diagram the state 'late', when a book is not returned by the specified deadline. This state is not strictly necessary, since whenever a book is not returned on time, the penalties can be associated with the reader and not with the book.

Exercise 8.4 Consider a system to manage a virtual bookshop, that includes the activities that go from the acquisition of the books up to their sale. The customers and the manager access the system through a web browser. The payments are made with a credit card. The bookshop works with new books acquired from publishers that have automatic acquisition systems. The system calculates the cost of the delivery based on the weight of the books and the delivery destination. The managers can produce reports about the bestsellers and the customers with more visits to the website. The system must also recommend books to the customers based on their previous interests. When a book is ordered, it is immediately sent if it is in stock; otherwise, it is requested from the publisher and the customer is informed about the expected delivery deadline.

- Construct a state diagram for the book entity. Consider the states 'available at catalogue', 'borrowed', 'in stock', 'sold', 'sent', and 'returned'.
- Construct a use case diagram, considering the participation of three different entities: book shop, customer, and credit card operator.
- Construct an activity diagram that models and relates the business processes of a book shop.

Glossary

Actor   Role that the users can have with respect to a given system, when interacting with it.

Ambiguity   Characteristics of the words, expressions, or sentences that express more than one possible understanding.

Analyst   Systems engineer specialised in the tasks related to the analysis phase; the same as requirements engineer.

Artefact   Tangible entity used or produced during the development process.

Candidate requirement   Requirement that was identified by some elicitation technique, whose incorporation in the system depends on the agreements that are established in the negotiation process.

Client   Entity that orders and pays for the development of a system, normally after negotiating the cost with the producer or provider.

Customer   Person that pays for acquiring a system, whenever it is developed and available to be used.

Derived requirement   Requirement that is obtained by refining a primary requirement.

Description   Representation of a descriptive model.

Developer   Professional that executes activities that contribute to the development and maintenance of a given technical system; the same as member of the development team.

Development   Lifecycle phases responsible for the construction of the system, including analysis, design, and implementation.

Diagram   Graphical representation of a collection of symbols in a two- or three-dimensional space.

Domain   Area of human knowledge or activity that is characterised by possessing a set of concepts and terms that the respective players know.

Embedded software   Software integrated in a system that cannot be classified as being of software.

Emergent property   Property that can be associated to the system as a whole, but not individually to each of its components.

© Springer International Publishing Switzerland 2016
J.M. Fernandes and R.J. Machado, *Requirements in Engineering Projects*,
Lecture Notes in Management and Industrial Engineering,
DOI 10.1007/978-3-319-18597-2

Engineering   Application of a systematic, disciplined and quantifiable approach to the analysis, design, implementation and exploitation of systems, using knowledge, principles, techniques and methods that originate from the empirical-scientific advances, in a ethical context to satisfy the needs of the human development.

Expert   Person that shows, in a given domain or subject, a deep knowledge, high-level skills and an extensive practical experience; the same as specialist.

Explicit requirement   Requirement that was requested by the clients and that is represented in the documentation.

Feature   Characteristics of a product or system clearly perceived as distinctive or differentiating by the users and that cohesively aggregates a set of functional and not functional requirements.

Firmware   Software embedded in electronic devices during their manufacturing.

Functional requirement   Functionality to be made available to the users of the system, partially characterising its behaviour as an answer to the stimulus that it is subject to.

Functionality   Capacity of a system to provide a useful function; concretisation in the solution final of what was identified as a functional requirement in the analysis phase.

Goal   High-level purpose (or intention), often related to the business or the organisation, that a stakeholder wishes to see reflected and fulfilled in the system under development.

Hardware   (of a computer) Electronic and electromechanical components, including, namely, the processor, the memory, and the input/output devices, that are part of a computer.

Implicit requirement   Requirement included by the development team, based on the domain knowledge that it possesses, in spite of not having been explicitly requested by the stakeholders.

Inspector   Someone who oversees or performs inspection over something.

Language   Countable set of strings over a fixed alphabet (set of characters).

Lifecycle   (of a system) Sequence of phases that starts when the system is mentally conceptualised and that ends when it is decommissioned.

Maintenance   Development activities occurring after the first release of an operational version.

Method   Set of generic guidelines that conduct the execution of a given set of activities, within the scope of a specific lifecycle phase.

Model   Abstraction of the system view, representing a perception or conceptualisation of that system made by the engineer.

Model of computation   Collection of rules that govern the semantics of the components and the communication among those components, within a given domain.

Necessity   Consequence of a problem that a person or an organisation has and that must be approached to justify the usefulness of the solution that satisfies it.

Negative stakeholder   Someone that desires that the system is not developed, neither put into operation.

Negotiation   Basic social process that is used with the objective of solving conflicts.

Non-functional requirement   Set of restrictions imposed to the system to be developed; the same as quality attribute.

Notation   Set of symbols/signs used for constructing representations.

Ontology   Explicit representation of a shared understanding of the important concepts in some domain of interest.

Paradigm   Way of organising knowledge; way of viewing the world.

Persona   Fictitious person that represents an important type of users of the product under development.

Phase   Abstraction of a set of activities.

Primary requirement   Requirement that originates directly from some stakeholder.

Process   Sequence of activities executed with the objective of achieving a given result.

Process model   Abstraction of a process, describing it in accordance with a given modelling perspective.

Product   Combination of (material and immaterial) goods and services that the supplier combines, in accordance with his commercial interests, to transfer established rights to the client.

Program   Text, written in a symbolic language capable of being interpreted by a computer, and constituted by a set of operations that operate over the data and that obey the controls that stipulate the execution moments.

Project   Temporary enterprise, composed of various coordinated activities with well-defined start and stop dates, undertaken to create a unique system.

Prototype   Easily modifiable and extensible working model of a system, in which only part of its aspects are considered, namely those that are related to the external view.

Questionnaire   Set of questions that serves for collecting information.

Requirement   Capacity that a system must possess, to satisfy the users necessities.

Requirements engineering   Set of activities that, in the context of a system development through an engineering project, permits eliciting, negotiating, and documenting the functionalities and the restrictions of that system.

Reverse engineering   Process of analysing a given system to represent it at a higher level of abstraction.

Risk   Potentially adverse circumstance that can have negative or perverse effects in the development process and in the final quality of the system.

Scenario   Story that describes the functional behaviour of a system and portrays a specific sequence of actions and events necessary for its execution.

Service   Coherent set of related functionalities, that allow users to satisfy a given necessity or objective.

Software   Programs, which include instructions, that are executed by the computer hardware, as well as the data that are operated by those instructions.

Software application   Software product developed to support the realisation of the individual tasks of the persons and the execution of the organisational processes.

Software engineering   Application of a systematic, disciplined and quantifiable approach in the context of the planning, development and exploration of software systems; application of engineering to the software domain.

Software process   Sequence of steps needed to develop or maintain software.

Software producer   Organisation or company that develops software systems.

Software product   Software system produced to be commercialised for or made available to the public in general; homogeneous software product, composed of programs, data structures, and documentation.

Specification   Formal and declarative representation of a prescriptive model.

Stakeholder   Person, group of persons or organisation that has some type of legitimate interest in a given system.

State   Ontological condition that persists for a significant period of time that is distinguishable and disjoint from other similar conditions.

Survey   Technique adopted in various domains that uses a questionnaire to collect and handle the information gathered from multiple respondents.

System   Identifiable and coherent set of components that interact to achieve a given objective.

System requirement   Detailed specification of a requirement, being generally a desirably formal model of the system.

System software   Set of programs that interact in an intense and direct way with the computer hardware.

Systems engineering   Interdisciplinary branch of engineering that is devoted to developing and managing complex and heterogeneous systems, according to a systematic approach.

Tailor-made software system   Software system developed by request of a given client for satisfying his own necessities and expectations; the same as custom software system or bespoke software system.

Traceability   Capacity that permits recording and following the requirements lifecycle, both upstream and downstream.

Use case   Set of actions, executed by the actors and by the system, so that a valuable result is produced to the users.

User   Any person that operates and interacts directly with the system, whenever it is in effective operation; the same as end user.

User requirement   Functionality that the system is expected to provide to its users or a restriction that is applicable to the operation of that system.

References

Abbott, R.J. 1983. Program design by informal descriptions. *Communications of the ACM* 26(11): 882–894. doi:10.1145/182.358441.

Abran, A., J.W. Moore, P. Bourque, and R. Dupuis. 2004. *Guide to the software engineering body of knowledge: 2004 edition–SWEBOK*. New York: IEEE Computer Society Press.

ACM/IEEE-CS (2004). Software Engineering 2004—curriculum guidelines for undergraduate degree programs in software engineering. ACM/IEEE-CS Joint Task Force on Computing Curricula. A Volume of the Computing Curricula Series.

Aczel, A. 1995. *Statistics: Concepts and applications*. Illinois: Richard D Irwin.

Adlin, T., and J. Pruitt. 2010. *The essential persona lifecycle: Your guide to building and using personas*. Burligton: Morgan Kaufmann.

Aho, A.V., M.S. Lam, R. Sethi, and J.D. Ullman. 2007. *Compilers: Principles, techniques, and tools*, 2nd ed. Boston: Addison-Wesley.

Albin, S.T. 2003. *The art of software architecture: Design methods and techniques*. New York: Wiley.

Alexander, I.F., and L. Beus-Dukic. 2009. *Discovering requirements: How to specify products and services*. Chichester: Wiley.

Alexander, I.F., and N. Maiden. 2004. *Scenarios, stories, use cases: Through the systems development life-cycle*. Chichester: Wiley.

Alexander, I.F., and R. Stevens. 2002. *Writing better requirements*. Harlow: Addison-Wesley.

Alperson, P. 1994. Introduction: The philosophy of music. In *What is music?: An introduction to the philosophy of music*, ed. P. Alperson. University Park: Penn State Press.

Arnold, R., and S. Bohner. 1996. *Software change impact analysis*. Los Alamitos: Wiley.

Aurum, A., and C. Wohlin. 2005a. *Engineering and managing software requirements*. Berlin: Springer. doi:10.1007/3-540-28244-0.

Aurum, A., and C. Wohlin. 2005b. Requirements engineering: Setting the context. In Aurum and Wohlin (2005a), pp. 1–15. doi:10.1007/3-540-28244-0_1.

Ballejos, L.C., and J.M. Montagna. 2008. Method for stakeholder identification in interorganizational environments. *Requirements Engineering* 13(4): 281–297. doi:10.1007/s00766-008-0069-1.

Bass, L., P. Clements, and R. Kazman. 1998. *Software architecture in practice*. Boston: Addison-Wesley.

Baumeister, R.F. 2005. *The cultural animal: Human nature, meaning, and social life*. New York: Oxford University Press.

© Springer International Publishing Switzerland 2016

J.M. Fernandes and R.J. Machado, *Requirements in Engineering Projects*,

Lecture Notes in Management and Industrial Engineering,

DOI 10.1007/978-3-319-18597-2

Bazerman, M.H., and D. Moore. 2012. *Judgment in managerial decision making*, 8th ed. Hoboken: Wiley.

Beck, K. 2000. *Extreme programming explained: Embrace change*, 1st ed. Boston: Addison-Wesley.

Beck, K., M. Beedle, A. van Bennekum, A. Cockburn, W. Cunningham, M. Fowler, J. Grenning, J. Highsmith, A. Hunt, R. Jeffries, J. Kern, B. Marick, R. C. Martin, S. Mellor, K. Schwaber, J. Sutherland, and D. Thomas. 2001. Manifesto for agile software development. www.agilemanifesto.org.

Beizer, B. 2000. Software is different. *Annals of Software Engineering* 10(1–4): 293–310. doi:10.1023/A:1018999919169.

Benyon, D. 2010. *Designing interactive systems: A comprehensive guide to HCI and interaction design*, 2nd ed. Harlow: Pearson.

Berander, P., and A. Andrews. 2005. Requirements prioritization. In Aurum and Wohlin (2005a), 69–94. doi:10.1007/3-540-28244-0_4.

Bernsen, N.O., and L. Dybkjær. 2009. *Multimodal usability*. London: Springer.

Berry, D.M. 2008. Ambiguity in natural language requirements documents. In Innovations for requirement analysis. From stakeholders needs to formal designs, vol. 5320, ed. B. Paech, and C. Martell, 1–7, Lecture Notes in Computer Science Berlin: Springer. doi:10.1007/978-3-540-89778-1_1.

Beyer, H.R., and K. Holtzblatt. 1995. Apprenticing with the customer. *Communications of the ACM* 38(5): 45–52. doi:10.1145/203356.203365.

Bjørner, D. 2001. *Software engineering*. Berlin: Springer.

Blundell, B.G. 2008. *Computer hardware*. London: Thomson.

Boehm, B.W. 1988. A spiral model of software development and enhancement. *IEEE Computer* 21(5): 61–72. doi:10.1109/2.59.

Boehm, B.W., C. Abts, A.W. Brown, S. Chulani, B.K. Clark, E. Horowitz, R. Madachy, D.J. Reifer, and B. Steece. 2000. *Software cost estimation with COCOMO II*. Upper Saddle River: Prentice-Hall.

Boehm, B.W., J.R. Brown, H. Kaspar, M. Lipow, G.J. Macleod, and M.J. Merrit. 1978. *Characteristics of software quality*. Amsterdam: North Holland.

Boehm, B.W., P. Grünbacher, and R.O. Briggs. 2001. Developing groupware for requirements negotiation: lessons learned. IEEE Software 18(3): 46–55. 2007, doi:10.1109/52.922725. Also available. In: Selby, R.W. Software engineering: Barry W. Boehm's lifetime contributions to software development, management, and research, John Wiley & Sons, Hoboken, NJ, USA.

Booch, G., R.A. Maksimchuk, M.W. Engle, J. Conallen, and K.A. Houston. 2007. *Object-oriented analysis and design with applications*, 3rd ed. Redwood City: Addison-Wesley.

Booch, G., J. Rumbaugh, and I. Jacobson. 1999. *The unified modeling language user guide*. Boston: Addison-Wesley.

Bosch, J. 2000. *Design and use of software architectures: Adopting and evolving a product-line approach*. Harlow: Addison-Wesley.

Bourque, P., R. Dupuis, A. Abran, J.W. Moore, and L. Tripp. 1999. The guide to the software engineering body of knowledge. *IEEE Software* 16(6): 35–44. doi:10.1109/52.805471.

Box, G.E.P., and N.R. Draper. 1987. *Empirical model-building and response surfaces*. New York: Wiley.

Brace, I. 2013. *Questionnaire design: How to plan, structure and write survey material for effective market research*. London: Kogan Page.

Brooks Jr, F.P. 1987. No silver bullet: essence and accidents of software engineering. *Computer* 20(4): 10–19. doi:10.1109/MC.1987.1663532.

Broy, M., and E. Denert (eds.). 2002. *Software pioneers: Contributions to software engineering*. New York: Springer.

Budd, T. 1997. *An introduction to object-oriented programming*, 2nd ed. Boston: Addison-Wesley.

Budgen, D. 1994. *Software design*. Harlow: Addison-Wesley.

Budinski, K.G. 2001. *Engineers' guide to technical writing*. Materials Park: ASM International.

Buxmann, P., H. Diefenbach, and T. Hess. 2013. *Software industry: Economic principles, strategies, perspectives*. Berlin: Springer.

Campbell-Kelly, M. 1995. Development and structure of the international software industry: 1950–1990. *Business and Economic History* 24: 73–110.

Campbell-Kelly, M. 2003. *From airline reservations to Sonic the Hedgehog: A history of the software industry*. History of Computing. London: MIT Press.

Ceruzzi, P.E. 1998. *A history of modern computing*. Cambridge: MIT Press.

Chandler, D. 2007. *Semiotics: The basics*, 2nd ed. London: Routledge.

Chemuturi, M. 2012. *Requirements engineering and management for software development projects*. New York: Springer.

Chen, P.P.-S. 1976. The entity-relationship model-toward a unified view of data. *ACM Transactions on Database Systems* 1(1): 9–36. doi:10.1145/320434.320440.

Chikofsky, E.J., and J.H. Cross II. 1990. Reverse engineering and design recovery: A taxonomy. *IEEE Software* 7(1): 13–17. doi:10.1109/52.43044.

Chung, L., and J.C.S. Prado Leite. 2009. On non-functional requirements in software engineering. In *Conceptual modeling: Foundations and applications, volume 5600 of Lecture Notes in Computer Science*, ed. A.T. Borgida, V.K. Chaudhri, P. Giorgini, E.S. Yu, et al., 363–379. Berlin: Springer. doi:10.1007/978-3-642-02463-4_19.

Clare, C.R. 1973. *Design logic systems using state machines*. New York: McGraw-Hill.

Classen, A., P. Heymans, and P.-Y. Schobbens. 2008. What's in a feature: A requirements engineering perspective. In *11th international conference on fundamental approaches to software engineering (FASE 2008), held as part of ETAPS 2008, volume 4961 of Lecture Notes in Computer Science*, ed. J. Fiadeiro and Inverardi, P., 16–30. Berlin: Springer. doi:10.1007/978-3-540-78743-3_2.

Coad, P., and E. Yourdon. 1991. *Object-oriented analysis*, 2nd ed. Englewood Cliffs: Prentice-Hall.

Cockburn, A. 2000. *Writing effective use cases*. New York: Addison-Wesley.

Cohn, M. 2004. *User stories applied for agile software development*, 2nd ed. Boston: Addison-Wesley.

Connell, J., and L. Shafer. 1989. *Structured rapid prototyping: An evolutionary approach to software development*. Englewood Cliffs: Yourdon Press.

Constantine, L.L., and L.A.D. Lockwood. 1999. *Software for use: A practical guide to the models and methods of usage-centered design*. Boston: Addison-Wesley.

Cook, N. 1993. *Beethoven symphony no. 9*. Cambridge: Cambridge University Press.

Cooper, A. 1999. *The inmates are running the asylum: Why high tech products drive us crazy and how to restore the sanity*. Indianapolis: Macmillan Publishing.

Cusumano, M.A. 2004. *The business of software: What every manager, programmer, and entrepreneur must know in good times and bad*. New York: Free Press.

Czarnecki, K., and U.W. Eisenecker. 2000. *Generative programming: Methods, tools, and applications*. New York: Addison-Wesley.

Dahlstedt, Å.G., and A. Persson. 2005. Requirements interdependencies: state of the art and future challenges. In Aurum and Wohlin (2005a), 95–116. doi:10.1007/3-540-28244-0_5.

Dardenne, A., A. van Lamsweerde, and S. Fickas. 1993. Goal-directed requirements acquisition. *Science of Computer Programming* 20(1–2): 3–50. doi:10.1016/0167-6423(93)90021-G.

Davies, C.G., and P.J. Layzell. 1993. *The Jackson approach to system development: An introduction*. Lund: Chartwell-Bratt.

Davis, A.M. 1990. *Software requirements: Analysis and specification*. Englewood Cliffs: Prentice-Hall.

Davis, A.M. 2005. *Just enough requirements management: Where software development meets marketing*. New York: Dorset House.

Davis, W.S. 1983. Systems analysis and design: A structured approach. Reading: Addison-Wesley.

DeMarco, T. 1978. *Structured analysis and systems specification*. New York: Yourdon Press.

Dennis, A., B. Wixom, and R. Roth. 2009. *Systems analysis and design*, 4th ed. Hoboken: Wiley.

Dijkstra, E.W. 1968. Letters to the editor: Go to statement considered harmful. *Communications of the ACM* 11(3): 147–148. doi:10.1145/362929.362947.

Dori, D. 2002. *Object-process methodology: A holistic systems paradigm*. Secaucus: Springer.

Douglass, B. 2004. *Real time UML: Advances in the UML for real-time systems*, 3rd ed. Boston: Addison-Wesley.

Ebert, C., and C. Jones. 2009. Embedded software: Facts, figures, and future. *IEEE Computer* 42(4): 42–52. doi:10.1109/MC.2009.118.

Erickson, J., and K. Siau. 2007. Theoretical and practical complexity of modeling methods. *Communications of the ACM* 50(8): 46–51. doi:10.1145/1278201.1278205.

Eriksson, H., M. Penker, B. Lyons, and D. Fado. 2004. *UML 2 toolkit*. Indianapolis: Wiley.

Fairbanks, G. 2010. *Just enough software architecture: A risk-driven approach*. Boulder: Marshall & Brainerd.

Fernandes, J.M., R.J. Machado, and S.B. Seidman. 2009. A requirements engineering and management training course for software development professionals. In *22th IEEE-CS conference on software engineering education & training (CSEE&T 2009)*, 20–5. IEEE CS Press. doi:10.1109/CSEET.2009.24.

Figueira, J., S. Greco, and M. Ehrgott. 2005. *Multiple criteria decision analysis: State of the art surveys*. Boston: Springer.

Fisher, R., and D. Shapiro. 2005. *Beyond reason: Using emotions as you negotiate*. New York: Penguin Books.

Fisher, R., W.L. Ury, and B. Patton. 1999. *Getting to yes: Negotiating an agreement without giving in*, 2nd ed. London: Random House.

Fleisher, W., and N. Gordon. 2010. *Effective interviewing and interrogation techniques*. Burlington: Academic Press.

Floyd, R.W. 1979. The paradigms of programming. *Communications of the ACM* 22(8): 455–460. doi:10.1145/359138.359140.

Forman, E.H., and S.I. Gass. 2001. The analytic hierarchy process–an exposition. *Operations Research* 49(4): 469–487. doi:10.1287/opre.49.4.469.11231.

Fowler, M. 2000. *Refactoring: Improving the design of existing code*. Boston: Addison-Wesley.

Fowler, M. 2004. *UML distilled: A brief guide to the standard object modeling language*, 3rd ed. Boston: Addison-Wesley.

Gajski, D.D., F. Vahid, S. Narayan, and J. Gong. 1994. *Specification and design of embedded systems*. Upper Saddle River: Prentice-Hall.

Galler, B.A. 1962. Definition of software. *Communications of the ACM* 5(1): 6. doi:10.1145/366243.366276.

Gane, C., and T. Sarson. 1979. *Structured systems analysis: Tools and techniques*. Englewood Cliffs: Prentice-Hall.

Gause, D.C., and G.M. Weinberg. 1989. *Exploring requirements: Quality before design*. New York: Dorset House.

Gasevic, D., D. Djuric, and V. Devedzic. 2006. *Model driven architecture and ontology development*, 3rd ed. Berlin: Springer.

Ghezzi, C., M. Jazayeri, and D. Mandrioli. 1991. *Fundamentals of software engineering*. Upper Saddle River: Prentice-Hall.

Gilb, T. 1997. Towards the engineering of requirements. *Requirements Engineering* 2(3): 165–169. doi:10.1007/BF02802774.

Gillham, B. 2008. *Developing a questionnaire*, 2nd ed. London: Continuum.

Gleick, J. 2011. *The information: A history, a theory, a flood*. New York: Pantheon Books.

Goguen, J.A., and C. Linde. 1993. Techniques for requirements elicitation. In: *International symposium on requirements engineering (RE 1993)*, 152–164. IEEE Computer Society Press. doi:10.1109/ISRE.1993.324822.

Goodwin, K. 2009. *Designing for the digital age: How to create human-centered products and services*. Indianapolis: Wiley.

Gottesdiener, E. 2002. *Requirements by collaboration: Workshops for defining needs.* Boston: Addison-Wesley.

Grady, R.B., and D.L. Caswell. 1987. Software metrics: Establishing a company-wide program. Upper Saddle River: Prentice-Hall.

Gregory, J., M. Aanestad, and S. Finken. 2013. Assignments. Qualitative research methods for course INF5220. Departament of Informatica, Universidade de Oslo, Noruega.

Grünbacher, P., and N. Seyff. 2005. Requirements negotiation. In Aurum and Wohlin (2005a), 143–162. doi:10.1007/3-540-28244-0_7.

Guizzardi, G. 2007. On ontology, ontologies, conceptualizations, modeling languages, and (meta)models. In *Databases and information systems IV*, ed. O. Vasilecas, J. Eder, and A. Caplinskas, 18–39., Frontiers in Artificial Intelligence and Applications Amsterdam: IOS Press.

Génova, G., M.C. Valiente, and J. Nubiola. 2005. A semiotic approach to UML models. *In CAiSE Workshops* 2: 547–557.

Harel, D. 1987. Statecharts: A visual formalism for complex systems. *Science of Computer Programming* 8(3): 231–274. doi:10.1016/0167-6423(87)90035-9.

Harel, D. 1992. Biting the silver bullet: Toward a brighter future for system development. *IEEE Computer* 25(1): 8–20. doi:10.1109/2.161283.

Harel, D., and E. Gery. 1997. Executable object modeling with statecharts. *IEEE Computer* 30(7): 31–42. doi:10.1109/2.596624.

Harel, D., and M. Politi. 1998. *Modeling reactive systems with statecharts: The STATEMATE approach.* New York: McGraw-Hill.

Hatley, D.J., and I.A. Pirbhai. 1987. *Strategies for real-time system specification.* New York: Dorset House.

Holder, R.D. 1990. Some comment on the analytic hierarchy process. *Journal of the Operational Research Society* 41(11): 1073–1076.

Horkoff, J., and E.S. Yu. 2011. Analyzing goal models: Different approaches and how to choose among them. In *ACM Symposium on Applied Computing 2011 (SAC 2011)*, 675–682. doi:10.1145/1982185.1982334.

Hull, M.E.C., K. Jackson, and J. Dick. 2011. *Requirements engineering*, 3rd ed. London: Springer.

Humphrey, W.S. 1995. A discipline for software engineering. Reading: Addison-Wesley.

Humphrey, W.S. 2005. *PSP: A self-improvement process for software engineers*, 1st ed. Upper Saddle River: Addison-Wesley.

Hutter, K., and K. Jöhnk. 2004. *Continuum methods of physical modeling: Continuum mechanics, dimensional analysis, turbulence.* Berlin: Springer.

IEEE (2014). SWEBOK 3.0–guide to the software engineering body of knowledge. IEEE Computer Society Press. Ballot version.

IIBA. 2007. A guide to the business analysis body of knowledge (BABOK guide), version 2.0. Toronto: International Institute of Business Analysis (IIBA).

Iverson, K.E. 1980. Notation as a tool of thought. *Communications of the ACM* 23(8): 444–465. doi:10.1145/358896.358899.

Jackson, D. 2006. *Software abstractions: Logic, language, and analysis.* Cambridge: MIT Press.

Jackson, M.A. 2001. *Problem frames: Analysing and structuring software development problems.* Boston: Addison-Wesley.

Jackson, M.A. 2002. Some basic tenets of description. *Software and System Modeling* 1(1): 5–9. doi:10.1007/s10270-002-0005-7.

Jacobson, I., M. Christerson, P. Jonsson, and G. Övergaard. 1992. *Object-oriented software engineering: A use case driven approach.* Wokingham: Addison-Wesley.

Jensen, K., and L.M. Kristensen. 2009. *Coloured Petri nets: Modelling and validation of concurrent systems.* Berlin: Springer.

Jönsson, P., and M. Lindvall. 2005. Impact analysis. In Aurum and Wohlin (2005a), 117–142. doi:10.1007/3-540-28244-0_6.

Jørgensen, J.B., S. Tjell, and J.M. Fernandes. 2009. Formal requirements modelling with executable use cases and coloured Petri nets. *Innovations in Systems and Software Engineering* 5(1): 13–25. doi:10.1007/s11334-009-0075-6.

Jureta, I.J., S. Faulkner, and P.-Y. Schobbens. 2006. A more expressive softgoal conceptualization for quality requirements analysis. In *25th international conference on conceptual modeling (ER 2006)*, 281–295. Springer. doi:10.1007/11901181_22.

Kaindl, H., and D. Svetinovic. 2010. On confusion between requirements and their representations. *Requirements Engineering* 15(3): 307–311. doi:10.1007/s00766-009-0095-7.

Kamsties, E. 2005. Understanding ambiguity in requirements engineering: setting the context. In Aurum and Wohlin (2005a), 245–266. doi:10.1007/3-540-28244-0_11.

Karlsson, J., and K. Ryan. 1997. A cost-value approach for prioritizing requirements. *IEEE Software* 14(5): 67–74. doi:10.1109/52.605933.

Karlsson, J., C. Wohlin, and B. Regnell. 1998. An evaluation of methods for prioritizing software requirements. *Information & Software Technology* 39(14–15): 939–947. doi:10.1016/S0950-5849(97)00053-0.

Karlsson, L., M. Höst, and B. Regnell. 2006. Evaluating the practical use of different measurement scales in requirements prioritisation. In *5th ACM-IEEE international symposium on empirical software engineering (ISESE 2006)*, 326–335. ACM. doi:10.1145/1159733.1159782.

Kim, J., M. Kim, and S. Park. 2006. Goal and scenario based domain requirements analysis environment. *Journal of Systems and Software* 79(7): 926–938. doi:10.1016/j.jss.2005.06.046.

Kit, E. 1995. *Software testing in the real world: Improving the process*. New York: ACM Press/Addison-Wesley.

Kittlaus, H.-B., and P.N. Clough. 2009. *Software product management and pricing: Key success factors for software organizations*, 1st ed. Berlin: Springer.

Knauss, E., and K. Schneider. 2012. Supporting learning organisations in writing better requirements documents based on heuristic critiques. In *Requirements engineering: foundation for software quality*, volume of 7195, Lecture Notes in Computer Science, ed. B. Regnell, and D. Damian, 165–171. Berlin: Springer. doi:10.1007/978-3-642-28714-5_14.

Kossiakoff, A., and W.N. Sweet. 2013. *Systems engineering: principles and practice*. Series in systems engineering and management Hoboken: Wiley.

Kovitz, B.L. 1999. *Practical software requirements: A manual of content and style*. Greenwich: Manning Publications.

Krutchen, P. 2003. *The Rational Unified Process: An introduction*, 3rd ed. Upper Saddle River: Addison-Wesley.

Kühne, T. 2005. What is a model? In *Language engineering for model-driven software development*, ed. Bézivin, J., Heckel, R., Dagstuhl Seminar Proceedings 04101, Internationales Begegnungs-und Forschungszentrum für Informatik, Schloss Dagstuhl, Germany.

Langford, J., and D. McDonagh. 2003. *Focus groups: Supporting effective product development*. London: Taylor & Francis.

Laplante, P. 2013. *Requirements engineering for software and systems*, 2nd ed. Boca Raton: CRC Press.

Laplante, P.A. 2007. *What every engineer should know about software engineering*. Boca Raton: CRC Press.

LaRocque, P. 2003. *The book on writing: The ultimate guide to writing well*. Arlington: Grey and Guvnor Press.

Lauesen, S. 2005. *Software requirements: Styles and techniques*. Harlow: Addison-Wesley.

Law, A.M., and W.D. Kelton. 1991. *Simulation modeling and analysis*, 3rd ed. Boston: McGraw-Hill.

Lee, E.A., and S.A. Seshia. 2011. Introduction to embedded systems-a cyber-physical systems approach. http://LeeSeshia.org.

Leffingwell, D., and D. Widrig. 2000. *Managing software requirements: A unified approach*. Boston: Addison-Wesley.

Lewicki, R.J., B. Barry, and D. Saunders. 2010. *Negotiation*. New York: McGraw-Hill/Irwin.

Lewicki, R.J., and J.A. Litterer. 1985. *Negotiation*. Washington: Richard D. Irwin.

Lidwell, W., K. Holden, and J. Butler. 2010. *Universal principles of design*. Beverly: Rockport.

Loucopoulos, P., and V. Karakostas. 1995. *System requirements engineering*. Englewood Cliffs: McGraw-Hill.

Ludewig, J. 2003. Models in software engineering. *Software and System Modeling* 2(1): 5–14. doi:10.1007/s10270-003-0020-3.

Luft, J. 1969. *Of human interaction*. Palo Alto: National Press Books.

Macaulay, L.A. 1996. *Requirements engineering*. London: Springer.

Machado, R., I. Ramos, and J.M. Fernandes. 2005. Specification of requirements models. In Aurum and Wohlin (2005a), 47–68. doi:10.1007/3-540-28244-0_3.

Machado, R.J., L.M. Alves, and P. Ribeiro. 2014. *Overcoming challenges in software engineering education: delivering non-technical knowledge and skills*, chapter Project-based learning: an environment to prepare IT students for an industry career, 230–249. IGI Global. doi:10.4018/978-1-4666-5800-4.ch012.

Marriott, K., and B.E. Meyer. 1998. Introduction. In *Visual language theory*, ed. K. Marriott, and B.E. Meyer, 1–4. New York: Springer.

Maslow, A.H. 1954. Motivation and personality. New York: Harper

Maslow, A.H. 1943. A theory of human motivation. *Psychological Review* 50(4): 370–396. doi:10.1037/h0054346.

Mehrtens, H. 2004. Mathematical models. In *Models: The third dimension of science*, ed. S. de Chadarevian, and N. Hopwood, 276–306. Stanford: Stanford University Press.

Meyer, B. 1988. *Object-oriented software construction*. Computer Science: Prentice-Hall.

Meyer, B. 2013. *Touch of class: Learning to program well with objects and contracts*. Berlin: Springer.

Monteiro, M.P., and J.M. Fernandes. 2006. Towards a catalogue of refactorings and code smells for AspectJ. In *Transactions on aspect-oriented software development I*, volume of 3880 *Lecture Notes in Computer Science*, ed. A. Rashid, and M. Aksit, 214–258. Berlin: Springer. doi:10.1007/11687061_7.

Moody, D. 2009. The "physics" of notations: toward a scientific basis for constructing visual notations in software engineering. *IEEE Transactions on Software Engineering* 35(6): 756–779. doi:10.1109/TSE.2009.67.

Moreira, A., R. Chitchyan, J. Araújo, and A. Rashid. 2013. *Aspect-oriented requirements engineering*. Berlin: Springer.

Morris, D., G. Evans, P. Green, and C. Theaker. 1996. *Object-oriented computer systems engineering*. London: Springer.

Muller, P.-A., F. Fondement, B. Baudry, and B. Combemale. 2012. Modeling modeling modeling. *Software and System Modeling* 11(3): 347–359. doi:10.1007/s10270-010-0172-x.

Murata, T. 1989. Petri nets: properties, analysis and applications. *Proceedings of the IEEE* 77(4): 541–580. doi:10.1109/5.24143.

Murphy, G.L. 2002. *The big book of concepts*. Cambridge: MIT Press.

Naveda, J.F., and S.B. Seidman. 2006. *IEEE Computer Society real-world software engineering problems: A self-study guide for today's software professional*. Hoboken: Wiley.

Norman, D.A. 1983. Some observations on mental models. In *Mental models*, ed. D. Gentner, and A.L. Stevens, 7–14. Hillsdale: Lawrence Erlbaum Associates.

Novak, J.D., and A.J. Cañas. 2008. The theory underlying concept maps and how to construct and use them. Technical report, IHMC CmapTools 2006–01, Rev 01–2008, Florida Institute for Human and Machine Cognition, Pensacola.

Osterwalder, A., and Y. Pigneur. 2010. *Business model generation*. Hoboken: Wiley.

Osterweil, L.J. 2007. A future for software engineering? In Future of software engineering (FOSE 2007), 1–11. *IEEE Computer Society*. doi:10.1109/FOSE.2007.1.

Page-Jones, M. 1980. *The practical guide to structured systems design*. New York: Yourdon Press.

Parnas, D.L. 1990. Education for computing professionals. *Computer* 23(1): 17–22. doi:10.1109/2.48796.

Parnas, D.L. 1999. Software engineering programs are not computer science programs. *IEEE Software* 16(6): 19–30. doi:10.1109/52.805469.

Parnas, D.L. 2010. Really rethinking 'formal methods'. *IEEE Computer* 43(1): 28–34. doi:10.1109/MC.2010.22.

Peck, M.E. 2012. The cryptoanarchists answer to cash. *IEEE Spectrum* 49(6): 50–56. doi:10.1109/MSPEC.2012.6203968.

Peirce, C.S. 1931–1958. Collected papers of Charles sanders peirce. Cambridge: Harvard University Press.

Perry, W. 2006. *Effective methods for software testing*, 3rd ed. New York: Wiley.

Peters, H. 2008. *Game theory: A multi-leveled approach*. Berlin: Springer.

Pfleeger, S.L., and J.M. Atlee. 2009. *Software engineering–theory and practice*, 4th ed. USA: Pearson Education.

Pohl, K. 1994. The three dimensions of requirements engineering: A framework and its applications. *Information Systems* 19(3): 243–258. doi:10.1016/0306-4379(94)90044-2.

Pohl, K. 1996. Process-centered requirements engineering. Taunton: Research Studies Press

Pohl, K. 2010. *Requirements engineering: Fundamentals, principles, and techniques*. Berlin: Springer.

Pressman, R.S. 2009. *Software engineering*, 7th ed. New York: McGraw-Hill.

Pretschner, A., M. Broy, I.H. Kruger, and T. Stauner. 2007. *Software engineering for automotive systems: A roadmap. In Future of software engineering* (FOSE, 2007), 55–71, Washington: IEEE Computer Society Press. doi:10.1109/FOSE.2007.22.

Purtova, N., E. Kosta, and B.J. Koops. 2015. Laws and regulations for digital health. In *Requirements Engineering for Digital Health*, ed. S.A. Fricker, C. Thümmler, A. Gavras, et al., 47–74. Switzerland: Springer, Cham. doi:10.1007/978-3-319-09798-5_3.

Pyster, A. (ed) 2009. Graduate software engineering 2009 (GSwE2009): Curriculum guidelines for graduate degree programs in software engineering. Stevens Institute of Technology. Version 1.0. ACM/IEEE-CS Joint Task Force on Computing Curricula. Final Report.

Rabiger, M., and M. Hurbis-Cherrier. 2013. *Directing: Film techniques and aesthetics*, 5th ed. Burlington: Focal Press.

Raiffa, H. 1982. *The art and science of negotiation*. Cambridge: Harvard University Press.

Richardson, J., T.C. Ormerod, and A. Shepherd. 1998. The role of task analysis in capturing requirements for interface design. *Interacting with Computers* 9(4): 367–384. doi:10.1016/S0953-5438(97)00036-2.

Ries, E. 2011. *The lean startup: How today's entrepreneurs use continuous innovation to create radically successful businesses*. New York: Crown Business.

Robertson, S. 2004. Scenarios in requirements discovery. In Alexander and Maiden, 39–59.

Robertson, S., and J.C. Robertson. 2006. *Mastering the requirements process*, 2nd ed. Upper Saddle River: Addison-Wesley.

Robinson, P.J. 1992. *Hierarchical object-oriented design*. Hertfordshire: Prentice-Hall International.

Rogers, Y., H. Sharp, and J. Preece. 2011. *Interaction design: Beyond human-computer interaction*, 3rd ed. Chichester: Wiley.

Rolland, C., C. Souveyet, and C.B. Achour. 1998. Guiding goal modeling using scenarios. *IEEE Transactions on Software Engineering* 24(12): 1055–1071. doi:10.1109/32.738339.

Roman, G.-C. 1985. A taxonomy of current issues in requirements engineering. *Computer* 18(4): 14–23. doi:10.1109/MC.1985.1662861.

Rossiter, J.R., and G.L. Lilien. 1994. New 'brainstorming' principles. *Australian Journal of Management* 19(1): 61–72.

Rowell, L. 1983. *Thinking about music: An introduction to the philosophy of music*. Amherst: University of Massachusetts Press.

Rumbaugh, J., M. Blaha, W. Premerlani, F. Eddy, and W. Lorensen. 1991. *Object-oriented modeling and design*. Englewood Cliffs: Prentice-Hall.

Rumbaugh, J., I. Jacobson, and G. Booch. 1999. The unified modeling language reference manual. Reading: Addison-Wesley.

Saaty, T.L. 1980. The analytic hierarchy process: Planning, priority setting, resource allocation. New York: McGraw-Hill

Salmre, I. 2005. *Writing mobile code: Essential software engineering for building mobile applications.* Boston: Addison-Wesley.

Schneider, G., and J.P. Winters. 1998. Applying use cases: A pratical guide. Object Technology. Boston: Addison-Wesley.

Schwaber, K., and M. Beedle. 2001. *Agile software development with Scrum,* 1st ed. Upper Saddle River: Prentice-Hall.

Seidewitz, E. 2003. What models mean. *IEEE Software* 20(5): 26–32. doi:10.1109/MS.2003. 1231147.

Seidl, M., M. Scholz, C. Huemer, and G. Kappel. 2014. *UML @ classroom: An introduction to object-oriented modeling.* Undergraduate Topics in Computer Science. Berlin: Springer.

Selic, B. 2003. The pragmatics of model-driven development. *IEEE Software* 20(5): 19–25. doi:10. 1109/MS.2003.1231146.

Selic, B. 2011. The theory and practice of modeling language design for model-based software engineering: A personal perspective. 3rd international summer school on generative and transformational techniques in software engineering III, 290–321. Berlin: Springer. doi:10.1007/978-3-642-18023-1_7.

Shaw, M. 1990. Prospects for an engineering discipline of software. *IEEE Software* 7(6): 15–24. doi:10.1109/52.60586.

Shaw, M. 2009. Continuing prospects for an engineering discipline of software. *IEEE Software* 26(6): 64–67. doi:10.1109/MS.2009.172.

Shelly, G., and H. Rosenblatt. 2011. *Systems analysis and design,* 9th ed. Boston: Course Technology.

Simon, M. 2007. *Storyboards: Motion in art,* 3rd ed. Burlington: Focal Press.

Simsion, G. 2007. *Data modeling: Theory and practice.* Bradley Beach: Technics Publications.

Smithin, J. 2000. What is money? introduction. In *What is money?*, ed. J. Smithin, 1–15. London: Routledge.

Sommerville, I. 2010. *Software engineering,* 9th ed. Boston: Addison-Wesley.

Stevens, P., and R. Pooley. 2006. *Using UML: Software engineering with objects and components,* 2nd ed., Object Technology Harlow: Addison-Wesley.

Stevens, R., P. Brook, K. Jackson, and S. Arnold. 1998. *Systems engineering: Coping with complexity.* Hertfordshire: Prentice Hall Europe.

Stevens, W.P., G.J. Myers, and L.L. Constantine. 1974. Structured design. *IBM Systems Journal* 13(2): 115–139. doi:10.1147/sj.132.0115.

Sutcliffe, A. 2002. *User-centred requirements engineering.* London: Springer.

Tanenbaum, A.S. 2006. *Structured computer organization,* 5th ed. Upper Saddle River: Prentice-Hall.

Thomas, K.W. 1976. Conflict and conflict management. In *Handbook in industrial and organizational psychology,* 889–935, ed. M.D. Dunnette. Rand McNally.

Thomé, B. 1993a. Definition and scope of systems engineering. In Thomé (1993b), 1–23.

Thomé, B., ed. 1993b. Systems engineering: Principles and practice of computer-based systems engineering. Chichester: Wiley

Trask, R.L. 1999. *Language: The basics,* 2nd ed. London: Routledge.

van Lamsweerde, A. 2000. Requirements engineering in the year 00: A research perspective. *22nd International conference on software engineering (ICSE 2000),* 5–19. New York: ACM. doi:10. 1145/337180.337184.

van Lamsweerde, A. 2009. *Requirements engineering: From system goals to UML models to software specifications.* Chichester: Wiley.

von Glasersfeld, E. 1995. A constructivist approach to teaching. In *Constructivism in education,* ed. L.P. Steffe, and J. Gale, 3–16. Hillsdale: Lawrence Erlbaum Associates.

Warfel, T. 2009. *Prototyping: A practitioner's guide*. Brooklyn: Rosenfeld Media.

Weinberg, G.M. 1993. *Quality software management, volume 2: First-order management*. New York: Dorset House.

Wexelblat, R.L. 1980. The consequences of one's first programming language. 3rd ACM SIGS-MALL symposium and the 1st SIGPC symposium on small systems, 52–55. New York: ACM. doi:10.1145/800088.802823.

Whytock, S. 1993. The development life-cycle. In Thomé (1993b), 81–96.

Wickelgren, I. 2012. Speaking science: why people don't hear what you say. *Scientific american*.

Wieringa, R.J. 1996. *Requirements engineering: Frameworks for understanding*. New York: Wiley.

Wieringa, R.J. 1998. A survey of structured and object-oriented software specification methods and techniques. *ACM Computing Surveys* 30(4): 459–527. doi:10.1145/299917.299919.

Wieringa, R.J. 2003. *Design methods for reactive systems: Yourdon, statemate, and the UML*. San Francisco: Morgan Kaufmann.

Xu, L., and S. Brinkkemper. 2007. Concepts of product software. *European Journal of Information Systems* 16(5): 531–541. doi:10.1057/palgrave.ejis.3000703.

Yang, H., A. Roeck, V. Gervasi, A. Willis, and B. Nuseibeh. 2011. Analysing anaphoric ambiguity in natural language requirements. *Requirements Engineering* 16(3): 163–189. doi:10.1007/s00766-011-0119-y.

Yourdon, E. 1988. *Managing the system life cycle: A software development methodology overview*, 2nd ed. Englewood Cliffs: Prentice-Hall.

Yu, E.S. 1997. Towards modeling and reasoning support for early-phase requirements engineering. In 3rd IEEE International Symposium on Requirements Engineering (RE 1997), pp. 226–235. IEEE Computer Society Press. doi:10.1109/ISRE.1997.566873.

Zave, P. 1984. The operational versus the conventional approach to software development. *Communications of the ACM* 27(2): 104–118. doi:10.1145/69610.357982.

Zave, P. 1997. Classification of research efforts in requirements engineering. *ACM Computing Surveys* 29(4): 315–321. doi:10.1145/267580.267581.

Zobel, J. 1997. *Writing for computer science: The art of effective communication*. Singapore: Springer.

Zowghi, D., and C. Coulin. 2005. Requirements elicitation: a survey of techniques, approaches, and tools. In Aurum and Wohlin (2005a), pp. 19–46. doi:10.1007/3-540-28244-0_2.

Index

© Springer International Publishing Switzerland 2016
J.M. Fernandes and R.J. Machado, *Requirements in Engineering Projects*,
Lecture Notes in Management and Industrial Engineering,
DOI 10.1007/978-3-319-18597-2

Printed in the United States
By Bookmasters